P9-DTQ-690

The Practical Handbook of

CONCRETE AND MASONRY

By Richard Day

Fawcett Publications, Inc.
1515 Broadway
New York, New York 10036

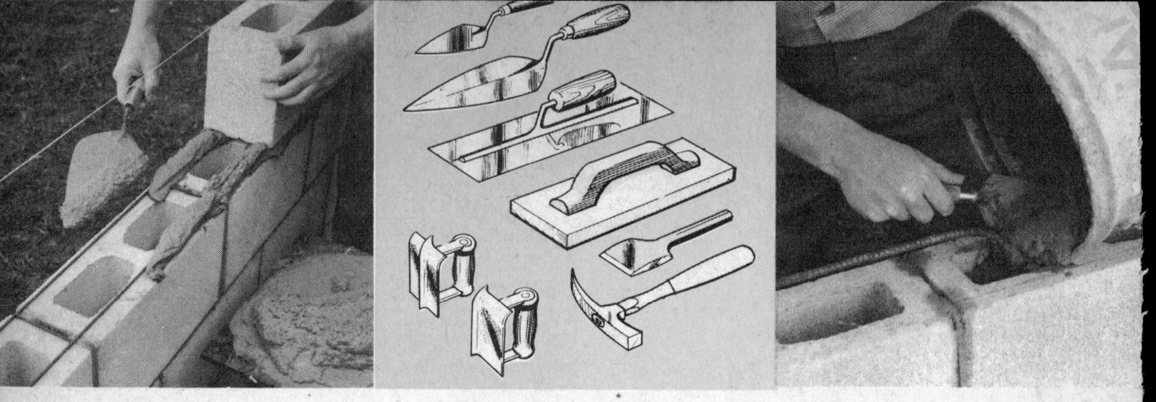

FRANK BOWERS: *Editor-in-Chief*

WILLIAM MIUCCIO: *Art Editor* • MICHAEL GAYNOR: *Asst. Art Editor*

Editorial Staff: DAN BLUE, ELLENE SAUNDERS, JOE CORONA, CAROL FRIEDMAN, COLLEEN KATZ, MARION LYONS

Art Staff: JACK LA CERRA, CARRIE SCHLIEPER, WENDY STONE, LARRY KEAVENY

JERRY CURCIO: *Production Manager*

ELLEN MEYER: *Production Editor*

How-To Art by Henry Clark
Cover Color by the Author

Printed in U.S.A. by
FAWCETT PRINTING CORPORATION
Rockville, Maryland

EIGHTH PRINTING

CONTENTS

HOW TO MAKE GOOD CONCRETE

Quality concrete is strong, durable, watertight and good-looking

Good concrete stays blemish-free for years as a monument to the person who made it.

Care in its preparation is the secret between quality concrete and poor concrete.
Robert Cleveland photo

All you have to do is look around you to see poor concrete work. It's everywhere. Poorly made concrete cracks, crazes, dusts, scales and spalls. It even can disintegrate completely.

Good concrete, on the other hand, suffers no such indecencies. It is a lasting monument to the person who built it.

There's always a reason for concrete failure. You can build poor concrete or you can build it good. Take your pick. The difference is in following a few simple directions. You'll find them in this chapter and the next. They are the essence of a 2-month Portland Cement Association training course on concrete technology. I took the course as a PCA employee. PCA is the one organization most interested in getting across the quality approach to concrete.

Before the course, I'd made every mistake in the book. The early concrete around my house shows them all. Now that I know the proper way to make and use concrete, it's a different story.

Concrete is a mixture of ingredients, much like a cake. Most important of these is portland cement. Portland cement is manufactured cement. It is not a brand. Some 70 firms in the U. S. and Canada make it from limestone, marl, shale and other rock products. The rocks are pulverized and fired in huge kilns. The resulting cement clinkers are ground into the fine-as-flour gray portland cement you can buy in bags. A little gypsum is ground in along with the cement to regulate setting time. Mixed with water, cement forms a paste that "glues" the other concrete ingredients together.

Concrete also contains fine aggregate (sand), coarse aggregate (gravel, stones or air-cooled slag), and water. These ags make up 66 to 78 percent of the volume of finished concrete. Concrete made with cement paste but no ags would be expensive. It would contain no cheaper materials to fill it out. It would also shrink a lot and crack as it set.

Too much cement paste is not good. Still there must be enough to surround every particle of aggregate.

Many people call concrete *cement*. They're not wrong, just square. If you

Portland Cement Assn.

Composite photo shows concrete sand (top) as it is delivered and (bottom) the sand when screened into its various particle sizes.

A good aggregate for concrete-making contains many particle sizes, ¼ inch and up.

Sliced concrete shows how sand and gravel fit together, surrounded by cement paste.

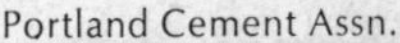

Portland Cement Assn.

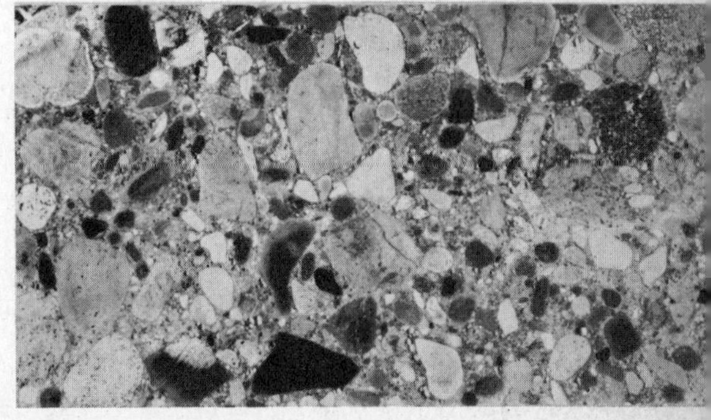

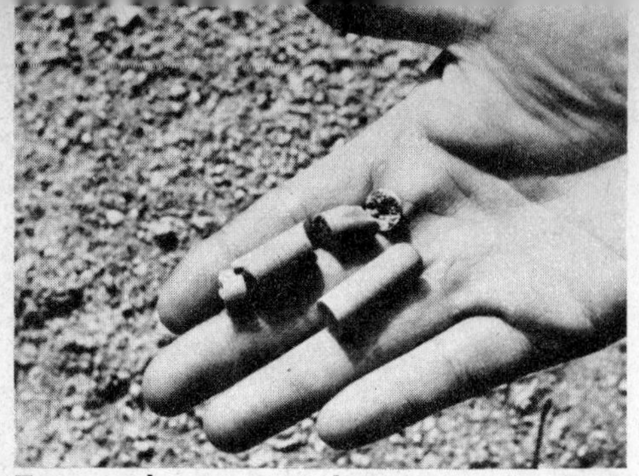

Too much water weakens concrete. Both plugs of hardened cement paste contain the same amount of cement but the lower column was mixed with less water, equal to 6 gallons a bag. It is strong. The other is weak, breaking because of too much water.

want to be with the in-crowd, call the gray powder *cement* and call the mixture made with cement *concrete*.

WHAT QUALITY CONCRETE IS

Quality concrete is ordinary concrete made by someone who cares. You can buy the separate ingredients and make it yourself, buy it prepackaged and mix it yourself or telephone your ready mix producer and order quality concrete delivered to you. Quality concrete is no more than a method for making strong, workable, durable, watertight, good looking concrete. Quality concrete handles well in the plastic state. In the hardened state it doesn't develop the faults you see in other concrete.

A concrete's quality depends largely on the binding qualities of the cement paste. Therefore, nearly every step in making quality concrete is aimed at getting a high-quality cement paste. The quality concrete method is easy to follow. It starts with the design of your project, its thickness, etc. Next comes design of the mix (richness, etc.). It also involves selection of materials, proportioning of them, mixing, forming, placing, finishing and jointing of the plastic concrete. Curing of the hardening concrete is the final step to quality. You have to go through most of these steps anyway. It takes very little additional time to do them right. Often you save time using the quality method.

Four numbers, 6666, are the key to quality concrete. They represent: cement, water, air and curing. Yes, there's air in good concrete. But more about it later.

CEMENT

The first "6" is for cement content. There should be 6 bags of cement per cubic yard of concrete. The section on proportioning will show you how to hit very close to the 6-bag figure when mixing your own concrete. Reputable ready packaged mixes contain close to this proportion of cement. Ready mix can be ordered to the 6-bag spec.

The second "6" stands for water content. There should be 6 gallons of water for each bag of cement in the mix, but no more. Whatever you do, don't put too much water in concrete. That's a no-no. Excess water is bad news. Nothing ruins it faster. While having lots of water makes it convenient to spread concrete, it knocks strength all to heck.

Concrete is a lot like gelatin dessert. Made with a little water, it produces a small amount of concentrated mix. Yum! But made with a lot of water, it makes a large amount of weak, insipid mix. Ech! Excess water evaporates, leaving pores and capillaries in the paste. These not

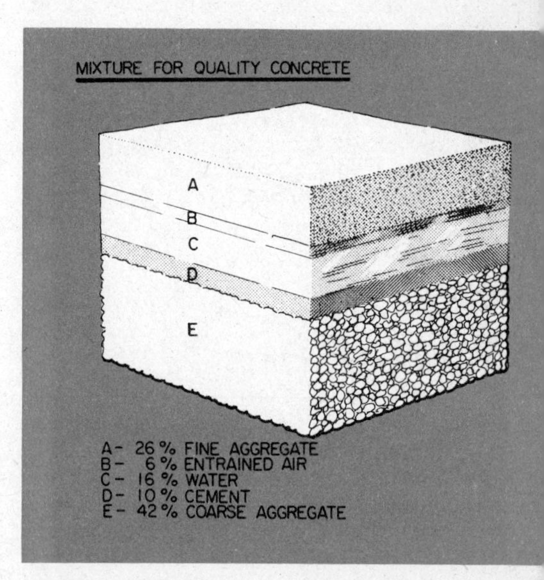

only weaken the paste, they make it porous and less durable. The work saved by flowing a soupy mix into place is later spent in finishing it. Then the extra water bleeds to the surface and gets in the way of your finishing operations.

Never never never use more water than 6 gallons per bag of cement. The table in this chapter will aid you.

AIR CONTENT

The third "6" is for air. There should be 6 percent entrained air in a quality mix. This is a special form of air that prevents freeze-thaw cycles from breaking up the surface of concrete. Air in the form of microscopic bubbles – hundreds of billions of them per cubic yard – is blended into the mix in either of two ways: (1) by using air-entraining cement or (2) by adding a separate air-entraining agent. This process is fully described later. When air-entrained concrete hardens, these bubbles stay on as cavities. Each cavity is a kind of "safety valve" where freezing water in the concrete can expand 9 percent or so without rupturing off chunks of the surface (scaling).

The resistance of air-entrained concrete to freezing and thawing is several hundred percent better than concrete containing no air. Air-entrained concrete

Two-year-old plain concrete (above) has scaled badly after repeated freezing and thawing. Air-entrained slab (below) is undamaged; is more resistant to frost.

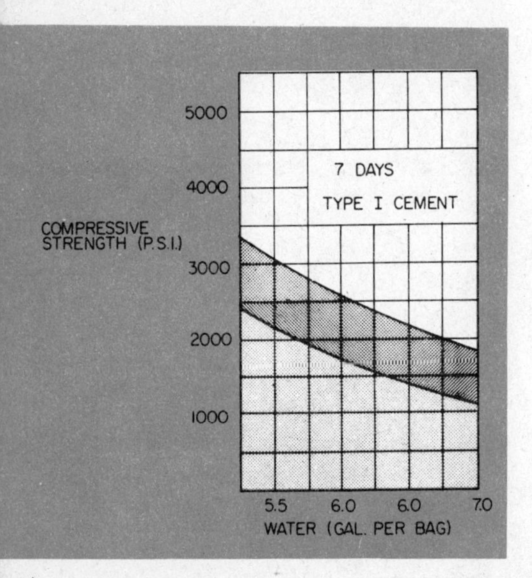

is a must for any project exposed to freezing. This includes patios, sidewalks, driveways, curbs, walls, planters and garage floors.

Indoors, and in nonfreezing climates, you may omit air-entrainment if you wish. But I recommend its use anyway. The tiny bubbles serve as ball bearings in the mix, lubricating it to increase workability. There isn't enough air to hurt strength. Air-entrained concrete is also cohesive, buttery. The disconnected air bubbles buoy up pieces of coarse aggregate keeping them from settling to the buttom of the mix and weakening your wall or slab.

What's more, the bubbles reduce what is called bleeding. Bleeding is when water separates from the cement paste and rises to the surface. Air-entrained concrete can be finished sooner than ordinary concrete.

WET CURE

The final "6" is for curing. Concrete should be wet-cured for at least 6 days. You can carry quality right up to the last and then blow it all by not curing properly.

Concrete starts gaining strength as soon as it begins to set. From this time throughout the concrete's whole life the tiny gels of cement are reacting chemically with water molecules in the mix. The process is called *hydration*. Never does every particle of cement react completely with every molecule of water, but the process goes rapidly at first. In about 6 days concrete gains most of the strength it needs. From then on hydration continues at a slower pace. The older concrete gets, the stronger it gets. Those first six days, in the presence of lots of water, get quality off to a good start.

Proper curing can develop up to 50 percent more strength in concrete. Curing methods are covered in the next chapter.

INGREDIENTS

Quality of the materials in it, and the way in which concrete is used, both have a bearing on the service your finished project will give.

Cement—All portland cement made in the United States and Canada is designed to meet rigid federal standards for quality. You needn't worry about it being good as long as the bag contains the statement, "Meets ASTM C150". Some imported portland cements may be good, too. See that they contain the ASTM C150 designation.

U. S.-made portland cement comes in 94-pound bags. Each bag contains 1 cubic foot of cement. On the bag the type of cement is designated. There are five types, designated by Roman numerals I through V. Type I is normal portland cement. It's the kind you should use. However, if you want your project to develop strength quickly, say in one to three days, you could use Type III high-early-strength portland cement. It may be tough to buy. Type I is usually the only kind you can get.

If the designation on the bag shows Type IA, it's air-entraining portland cement. If you use this type, you need not add an air-entraining agent to the mix. Another type of cement is white portland cement. Use it where you want a pure white color or for bright colored concrete using mineral pigments. Still another type of cement is masonry cement. Use it for mixing mortar, but never for making concrete. The lime it contains weakens the mix.

Plastic cements are widely sold in some parts of the country. Plasticizing agents have been added to ordinary Type I cement during manufacture. They're commonly used for making mortar, plaster and stucco, but not concrete. What you want is portland cement, Type I or Type IA, for most uses.

Portland cement that has been stored where moisture can get at it sometimes hardens in the bag. This differs from a condition called "warehouse pack," which is a stiffening of cement around the edges of the bag. Warehouse pack can be cured by rolling the bag on the floor. When it's used, cement should be free-flowing and free of lumps that can't be broken between your thumb and finger.

Water—Water that you can drink is nearly always suitable for making good concrete.

SAND AND STONES

Aggregates—Ags are termed *fine* or *coarse* according to their particle size. Sand, the fine aggregate, contains particles from dust size up to about ¼ inch. Gravel or stones, the coarse aggregate, should contain particles from ¼ inch up to a maximum specified size. Common

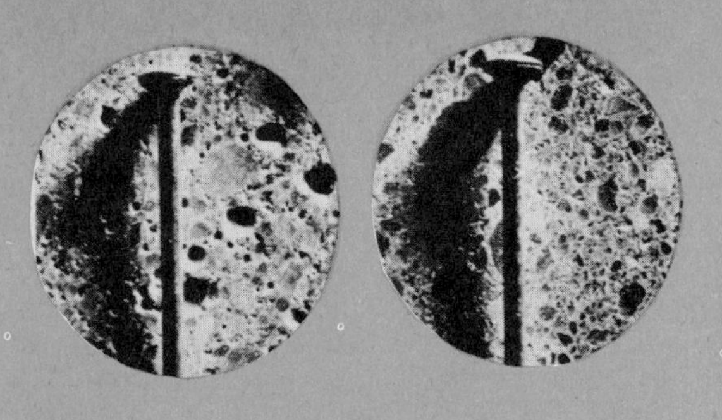

Entrained air bubbles, microscopic next to a pin, show up in greatly enlarged photo of air-entrained concrete (left) compared to plain concrete (right). The bubbles act as "relief valves" for frost pressures in winter.

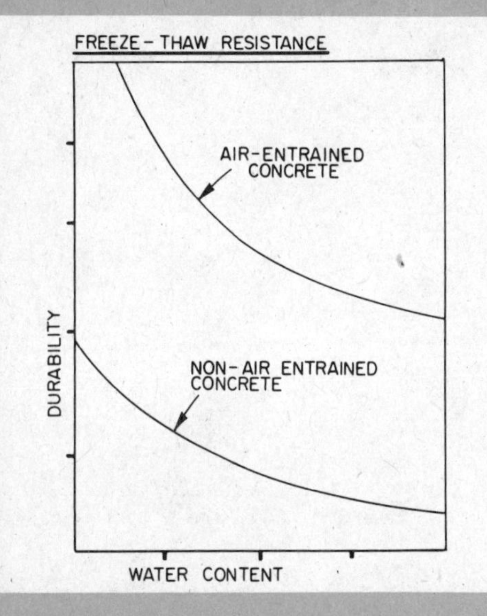

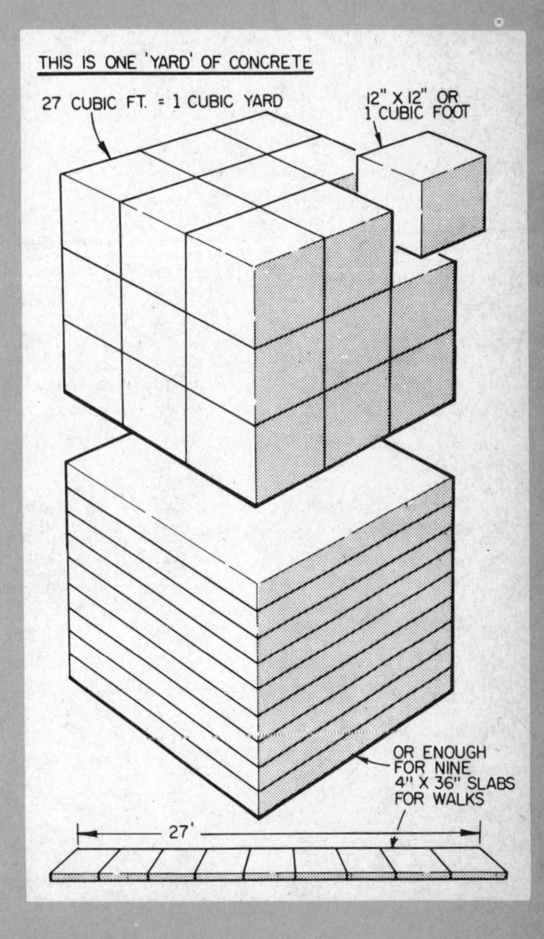

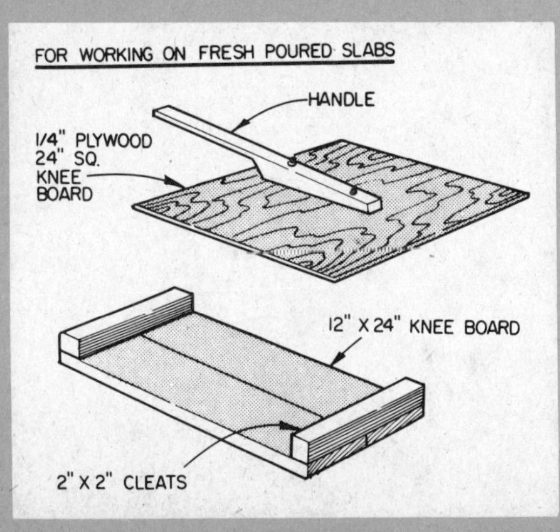

maximum sizes are ⅜, ½, ¾ and 1 inch. The largest size you can use depends on the project. The larger the ags, the less cement paste is needed per cubic yard. In general use a maximum size ag of about one-fifth the thickness of a wall or slab. For a 4-inch thick project, you'd use coarse aggregate up to ¾-inch in size.

For efficient use in concrete, the fine

Damp sand contains little water and falls apart after being squeezed in your hand.

Wet sand, when squeezed, forms a ball, yet leaves little moisture on your fingers.

Stony mix, harsh for finishing, contains too much coarse aggregate, not enough sand.

Workable mix is just right for easy finish. Spaces between all the stones are filled.

and coarse aggregates should be graded in steps from the smallest to the largest particle sizes. In other words the particles should not all be of one or two sizes. Big particles are needed to add bulk to the mix. The smaller ones fill the in-between spaces among the bigger pieces, and so on down to dust. It's like a mixture of watermelons, cantaloups, tomatoes, radishes, beans, peas, grape seeds and radish seeds all surrounded by glue. Not very taste-tempting, but a good range of sizes.

Very wet sand, will form a ball in your hand, but leaves moisture on your fingers.

Sandy mix has too much sand, not enough ag. Will finish easy but is uneconomical.

Concrete made from aggregates without a good range of sizes needs more of the high-cost cement paste to take the place of the missing sizes. Don't use fine sandbox or mortar sand in concrete. It doesn't have any of the larger particles needed.

Aggregates used in concrete must be able to withstand freezing and thawing, just like the concrete itself. They must also be strong and chemically stable. Avoid using aggregates that are soft, weak, laminated or break when squeezed hard. All ags should be free of fine dust, silt, loam, clay and vegetable matter. Keep them clean in the stockpiles. Cover with polyethylene sheets to ward off leaves and discourage cats. If you use unwashed aggregate from local pits or creeks, be sure it has been proven to make good concrete.

Buy your ags from a reputable source. I recommend a ready mix producer, Since he understands concrete. Not many other dealers do. You can use either gravel or crushed stone as a coarse ag. Pieces should not be long and thin. The more nearly rounded the better.

Additives—Things you put into concrete besides the cement, water and aggregates are called additives. Air-entraining agent is an additive. This is the only additive you should use. Others, such as calcium chloride, salt and flyash, are sometimes used by professionals for specific applications. You're better off without them. There's a good deal of disagreement as to how much good they do anyway.

Air-entraining agent is new to us home handymen. You've likely never seen any before. It's oily, soapy and dark. Looks like old crankcase oil. It works much like a detergent to make bubbles in the mix. A detergent, however, would make the wrong kind of bubbles. If you live in

Your concrete should look like two samples at left; the sample at right is too wet.

a cold climate, your ready mix dealer is sure to have air-entraining agent in bulk. Take him a gallon can and ask him to sell you some from his big drum. It isn't expensive. A gallon should last you ages.

If the brand your dealer has is *Darex*, you can use the quantities shown in the trial mix table in this chapter. This is figured at the rate of 1 ounce of *Darex* per bag of cement. *Darex* has a self-limiting feature. It can't "over-air" your mix. Some brands can. If your dealer uses another brand of agent, ask him to help you figure how much you should use to hit 6 percent entrained air.

MAKING QUALITY CONCRETE

Unless you order ready mix from a dealer, the job of making concrete is up to you. For all small jobs – anything less than a cubic yard – mix-it-yourself is the best method. Larger jobs can be broken down into a series of small jobs. A patio that's squared off into 4X4-or 5X5-foot sections, with form boards left in, can be cast section by section. A sidewalk or driveway can be built with keyed construction joints between one day's production and the next. This puts a big job

Surest way to get consistent quality is to measure out all the materials in similar sized batch cans before dumping into mixing drum.

Household measuring teaspoons and tablespoons are used to measure air-entrained agent accurately. Add liquid to the water.

in the mix-it-yourself class if that's what you want.

The big advantage of mixing your own concrete is that you can pace the job to suit yourself. There's no backbreaking rush to get a ready mix truck unloaded. And there's no huge amount of concrete dumped out on the grade that has to be placed and finished before it sets up. You can take things easier. Do as much as you feel like, then quit.

There *are* drawbacks. When you do the mixing, the job of choosing materials to go into concrete is yours. So is the responsibility for correctly proportioning the ingredients.

MACHINE-MIXING

There are two ways to mix your own concrete: by hand and with a machine. A big project calls for a concrete mixer. Machine-mixing saves you time and energy for placing and finishing. It also does a better job of combining ingredients. In one day with a decent-sized mixer, one man can handle mixing, placing and finishing of one to two cubic yards of concrete. If you'll have much use for it, buy a mixer. A number of jobs over the years will pay for a mixer. Like a good power tool, a concrete mixer can be a lifetime investment. You can help out on the cost by renting your mixer to neighbors. The large contractor-type mixer I bought used for $125 has served me and my neighbors for more than eight years. It's still worth about what I paid. The pnuematic tires make it fully portable. Maintenance is merely an occasional greasing of the fittings and a spark plug cleaning. But if you haven't much need for a mixer, it will pay you to rent one when you need it.

Concrete mixers you can use vary from tiny 5-gallon-pails to contractor-

Weigh trial batch materials on a scale using the quantities shown in the table, page 14, to establish your proportions accurately.

Once you get correct proportions, mark the batch can. Then you can fill them to the line each time, sure of correct proportions.

size 7-cubic-foot jobs. The smallest ones make a patio-stone-sized batch. A 7-cubic-footer will mix 4 cubic feet of concrete at one time. You can get either a gasoline-or electric-powered mixer. The gas-engine type gives more flexibility of mixer location. Otherwise you must locate where you can plug the power cord. Electric mixers generally are easier to maintain. Cost less, too.

A mixer's size is always given as the fluid capacity of the mixing drum. Actual mixing capacity is about 60 percent less. For capacity see the identification plate attached to the mixer.

Putting materials into a concrete mixer in the proper order helps keep them from sticking to the sides. (l) Pour about one-tenth of your premeasured water into the rotating drum. (2) Shovel in the coarse aggregate. (3) Shovel the sand into the drum. (4) Measure and dump in the cement. (5) Finally add the rest of the water.

Air-entraining agent is added to the water. Measure it with household measuring spoons. There are 2 tablespoons (tb.) or 6 teaspoons (tsp.) per ounce.

Concrete should mix for at least a minute after all the ingredients are in the drum. A 3-minute mixing time is preferable. The concrete should have a uniform color before discharging it. Rinse the mixing drum thoroughly after use by adding water and a few shovels of gravel while the drum is turning.

Don't overload a concrete mixer or it can't do a thorough job of combining ingredients. Size your batches to stay within the mixer's working capacity (see table).

HAND-MIXING

Prepackaged concrete mixes are a great convenience for small jobs. Any job that's small enough for hand-mixing is probably small enough to use a prepackaged mix. These mixes usually come in 25-, 45-and 90-pound bags. With a ready packaged mix the job of selecting materials and proportioning has already been done for you. All you do is add water, then mix. Follow the directions on the bag.

Prepackaged concrete mixes can produce excellent quality concrete. They're available in gravel-mix and sand-mix. Use gravel-mix for most concrete uses, sand-mix for thin slabs and small precast items.

The big drawback to prepacks is cost. On larger jobs they cost much more than ordering cement, sand and gravel and mixing your own. If you're in doubt, figure the cost both ways. Then decide whether the added convenience of a prepack is worth its extra cost.

With hand-mixing forget about air-entrainment. The mixing isn't vigorous enough to create the many air bubbles needed. Prepackaged mixes, therefore,

"6666" CONCRETE TRIAL MIXES

SIZE OF BATCH	CEMENT	WATER Damp Sand	WATER Wet Sand	WATER Very Wet Sand	SAND Damp Sand	SAND Wet Sand	SAND Very Wet Sand	STONE	AIR ENT. AGT. (DAREX)
1 cu.ft.	21 lb.	10 lb.	9 lb.	7½ lb.	46 lb.	47 lb.	49 lb.	63 lb.	2 tsp.
2½ cu.ft.	52 lb.	25½ lb.	22 lb.	19 lb.	116 lb.	118 lb.	122 lb.	157 lb.	1 tb.
3½ cu.ft.	73 lb.	35½ lb.	32½ lb.	26½ lb.	162 lb.	165 lb.	171 lb.	220 lb.	1½ tb.
5 cu.ft.	105 lb.	51 lb.	46 lb.	38 lb.	231 lb.	236 lb.	244 lb.	315 lb.	2 tb.
Write your own results here									

QUANTITIES TO ORDER
(allow 10 percent for aggregate waste)

Concrete Needed	Cement 1 bag = 94 lb.	Sand	Stone or Gravel
¼ cu. yd.	2 bags	350 lb.	500 lb.
½ cu. yd.	3 bags	700 lb.	1000 lb.
1 cu. yd.	6 bags	1400 lb.	2000 lb.

Locate your sand, stones and cement conveniently for batching into mixer drum.

Sand and stones, once measured, can then be shoveled into mixer by counting shovels.

If you place bags of cement in a wheelbarrow you won't be wasting spilled cement.

CUBIC YARDS OF CONCRETE IN SLABS

Area in square feet (length times width)	THICKNESS IN INCHES 4	5	6
50	0.62	0.77	0.93
100	1.2	1.5	1.9
200	2.5	3.1	3.7
300	3.7	4.6	5.6
400	5.0	6.2	7.4
500	6.2	7.7	9.3

Measure water accurately before dumping into mixer. Let stir at least one minute.

don't contain air-entraining agents. If you use them in a mixer, add the agent.

Hand-mix small batches in a wheelbarrow, on a platform or concrete slab. Dump the materials in a pile and mix them dry with a shovel or hoe. Make a depression in the center of the mix, add some of the water and mix more. eep adding premeasured water and mix until all the water has been put in and the mix is uniform in color and workable. Every piece of aggregate should be completely coated with cement paste.

QUANTITIES OF MATERIALS

It's easy to figure how much cement, sand and stone will be needed to build a project. Multiply the project's *length* in feet times its *width* in feet, time its *thickness* in inches. Then divide by 324. This gives the number of cubic yards needed. Suppose a wall 6'x4'x24' is to be built. Multiply 6 times 4 times 24 and divide by 324. You'd need 1-3/4 cubic yards of concrete to complete the wall. Next look at the table in this chapter for the correct quantities of materials to order. For instance, on a project that calls for 1-3/4 cubic yards of concrete, you'd add the amounts of cement, sand and coarse aggregate needed for 1 cubic yard, plus 1/2 cubic yard, plus 1/4 cubic yard. From the table this would be 11 bags of cement, 2450 pounds of sand, 3599 pounds of coarse aggregate. The table allows about 10 percent for aggregate stockpile waste.

TRIAL MIX

Because all aggregates are not graded alike, you must make a trial mix. A trial mix is one full batch of concrete proportioned, mixed, inspected and then adjusted, if necessary, for aggregate gradation. The quantities in the trial mix table are for an average-graded aggregate. Make your first trial mix using these proportions.

There's one problem to getting the right amount of water. Sand contains water, too. This has to be considered as water in the mix. Sand-held water combines with the cement. Therefore, you don't add six gallons of water per bag to end up with that amount of water. You add something less than this to allow for water that's already in the sand.

Give your sand the hand-squeeze test before you proportion a mix and use the quantities in the table according to the sand's wetness. The wetter the sand the less water you use in a mix. There's no problem to it. The proportion table tells you how much water to use with each of three wetnesses of sand. In each case you'll end up with the equivalent of 6 gallons of water per bag of cement.

To use the table, first decide what size batch you can mix. Then choose the weights opposite your batch-size as follows:

1. *Cement*—Use the figure given.

2. *Water*—The amount depends on the wetness of your sand. Test the sand and select the proper figure. If you'd rather measure gallons than pounds, 8.3 pounds of water make a gallon (U. S.).

3. *Sand*—The amount depends on its wetness. Select the proper figure based on your wetness test.

4. *Coarse aggregate*—Use the figure given.

5. *Air entraining agent*—Use the figure given (tsp. is *teaspoon;* tb. is *tablespoon*).

Mix proportions are given in pounds so you can weigh them out in batching pails on a bathroom scale. When you weigh out materials on a scale, don't forget to "zero" the scale with an empty batch pail on it. The weighing method is easier and more accurate than constructing a square-foot box or the like for accurately measuring materials.

Dump out a sample of trial mix. (see photos.) If the mix is too stiff to be worked well, make a second trial batch and use less sand and gravel.

If the mix is too wet, weigh out more sand and gravel and add it to the original trial batch and mix in. If the mix is too sandy, use less sand in a second trial batch. If too rocky, use less stones.

Never vary the water or cement content. Always adjust the mix by varying the amounts of sand and coarse aggregate. Keep weight records so you know just how much sand and coarse ag is enough. With practice you may get so expert that you can judge consistency and sand-stone balance while the concrete is mixing.

Once you have everything written down, the only variable is water in the sand. If the stockpile of sand dries out or gets rained on, you may need to adjust your mix proportions. Check the sand for wetness before you start work each day.

If your aggregate source varies, you'll have to make a new trial batch for each new combination of aggregates. No adjustments are necessary for cement and water, since they don't vary from place to place enough to affect concrete.

Subsequent batches can be made by using the same number of shovels of sand and coarse aggregate that it took to make the weights in an adjusted trial batch. A mark can be put on the water pail. The same with a dry pail for batching ags and cement. Then you're ready to go on batch after batch.

MORTAR

Mortar is a mixture of masonry cement, sand and water. Sometimes portland cement and lime are mixed with sand and water to make mortar. However, the use of masonry cement, or else the ready-packaged mortar mix to which only water need be added, are the simplest and best ways for you to make mortar.

To mix a batch of mortar add to one part of mortar cement from 2¼ to 3 parts of damp, loose masonry sand. Masonry sand is finer than sand for concrete and it produces a more workable mortar. Mix the ingredients dry first. Then add enough clean water to produce the workability you want. In mortar you should use *as much water as possible,* without destroying the workability.

Mix only enough mortar to be used in 2½ hours or less. Within this time, adding water and remixing are permissible to retemper the mortar. Mortar that is older than 2½ hours and has partially set should be thrown away and a new batch mixed.

Colored mortar may be made by adding a concrete coloring agent.

Mortar may be mixed, either by hand or by mixer. For hand-mixing, use a wheelbarrow and hoe. First spread the masonry cement on top of the sand and mix the two. Mix from both ends until the sand-cement mixture takes on a uniform color. Add about three-fourths of the water and mix until all of the materials are uniformly damp, then add more water in small amounts and continue mixing until you get satisfactory workability. Let the batch stand for 5 minutes, then remix it thoroughly without adding water.

Mortar is better if you make it in a concrete mixer. The small 5-gallon-pail type is ideal for mixing mortar. With it, one batch can be mixing while you are using the previously mixed batch. Such a mixer has numerous other uses for concrete and masonry jobs around the house.

Machine-mixed materials should be batched as follows: Mix three-fourths of the water, one-half of the sand and all of the cement first. Then add the rest of the sand and mix briefly. Add water, a little at a time, until the workability is the way you want it. After the last water has been added, mix the mortar at least 5 minutes. Completely empty the mixer drum before starting the next batch.

Once mixed, mortar is easiest handled on an 18-inch square piece of plywood. From there you can pick it up with the trowel as needed. In hot, dry weather, wet the mortarboard before putting mortar on it. This will help to retain the workability of the mortar for a longer time.

TOOLS FOR CONCRETE WORK

Experience proves adaptability of these tools for concrete work

The best wheelbarrow for concrete work is a sturdy, rubber-tired contractor's barrow.

Besides the mixer, batch pails and bathroom scales already mentioned, there are a few other tools you should have before starting a concrete job. None are fancy or costly. We'll take them in the order they're normally used.

Wheelbarrow—You can get by with an ordinary garden wheelbarrow to transport concrete from the mixer or ready mix truck to the jobsite. But you're better off with a rubber-tired contractor's wheelbarrow for several reasons. Garden barrows don't hold much concrete. They are tippy and tend to spill your mix over the side. A full load of concrete is too much weight for the garden variety. Worse yet, garden barrows with steel wheels shake the mix as it's being moved. This makes the coarse aggregate settle to the bottom, creating what is called segregation. A smooth ride on a rubber-tired wheelbarrow delivers the mix to the job in the best condition.

Shovel—While you can batch the mixer with almost any kind of shovel, you'll need a flat-ended shovel for working concrete. If you don't have one, you can make-do with another type. The work is likely to go harder, though.

Straightedge—Find an unwarped 2×4 long enough to reach across your forms and stick out about 6 inches on either side. This screed is see-sawed back and forth over the forms while slowly draw-

A 2 x 4 can be used as a straightedge for striking off concrete flush with the forms.

Make your own darby. Use it where quarters are too cramped to swing a bull float.

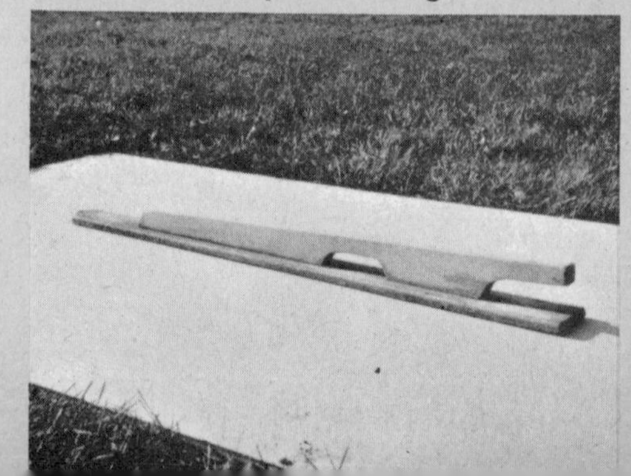

ing it forward. The action gently strikes off your freshly placed slab to proper grade. The maximum strike-off span should be 12 feet.

Bull Float—You can make a serviceable bull float out of a broom handle and flat piece of wood 8 inches wide and 4 feet long. A bull float lets you smooth a slab initially without ever setting foot on it. The float's long handle is raised when pulling the float and lowered when pushing it. That way the blade rides over the surface like a water ski.

Darby—A darby serves the same purpose as a bull float. It is more often used indoors where there isn't room for the bull float's long handle. A darby is harder to manage. You can make your own darby out of wood. Start with a long, flat piece about 3½ inches wide and 4 feet long. Fasten on an angling handle and you have it.

Both the bull float and darby eliminate high and low spots or ridges left by the straightedge after strikeoff. They also embed coarse aggregate for subsequent hand-floating and troweling.

Edger—Edgers are made in many sizes to put smooth curves around the

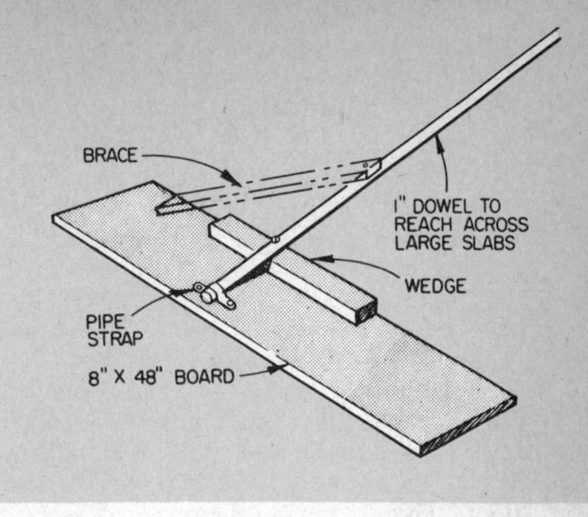

You and your helpers should have shovels. The flat-end type is best for this work.

You can use a bull float like a pro. Lower handle when pushing, raise when pulling.

Edgers are made in various sizes and designs to put rounded corners on concrete.

Bit depth on your jointer should be 1/5 the slab depth. A one inch does most jobs.

outsides of slabs. All are about 6 inches long but they vary in width from 1 1/2 to 4 inches. Lip lengths range from 1/8 to 5/8 inch. The radius may vary from a sharp 1/8-inch curvature to a gentle 1 1/2-inch. Edgers may have straight or curved-up ends. The curved-end toboggan type is easier to use, doesn't dig in. Edging a slab improves its looks and reduces the risk of edge breakage.

Jointer—Jointers are about 6 inches long and vary from 2 to 4 1/2 inches wide. They have shallow, medium or deep bits (cutting edges). These form joints from 3/4 to 2-3/16 inches deep. The cutting depth should be one-fifth the depth of your slab. A 4-inch-thick slab needs a ¾-inch-deep joint. A 1-inch jointer is most useful on the slabs you're likely to build. The jointer is used to cut contraction joints that predetermine the location of any possible cracks.

Hand Float—Hand floats are made of aluminum, magnesium or wood. The basic float is wood. You can make it yourself of a piece of wood 3½ to 4½ inches wide and about 15 inches long. Screw on a handle, recessing the screw heads in the float's bottom. A wood float is used to prepare the concrete surface for steel-troweling. It brings cement paste to the surface and forces down pieces of stone that stick out above the surface.

Metal floats are used on concrete that is to get something else than a troweled smooth finish. They're easier to work and leave the surface smoother than wood floats.

Special purpose floats are faced with cork, sponge rubber, canvas, etc. Each produces a different effect in the concrete surface. Usually a metal float finish is left as the final one. However, a metal float may be used before brooming the surface to create a nonslip, striated texture.

Steel Trowel—The mason's steel trowel puts a dense, smooth finish on concrete. Steel trowels come in many sizes. It's nice to have two, a large one and a small one. For the first troweling a trowel 4½ inches wide and 16 to 20 inches long is best. For the final troweling a smaller trowel is much easier to handle. It should be about 12X3 inches. If you use just one trowel, get it about 4X16 inches.

Kneeboards—Make a pair out of ⅜-inch exterior plywood with 1×2 or 2×2 cleats at the ends. Kneeboards let you get onto the concrete as soon as it's ready for final finishing. Your knees rest on one kneeboard, your toes rest on the other.

If you can reach all portions of a slab from outside the forms, kneeboards are not necessary.

Another tool, a small trowel, is useful in cleaning concrete off of wheelbarrows, mixers, straightedges and form boards. It's also great for finishing in tight spots

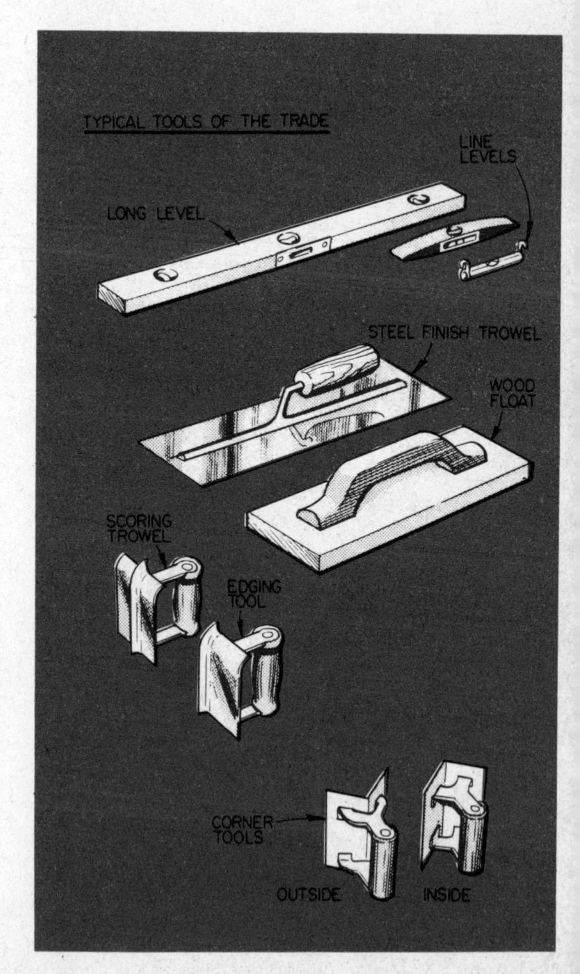

where a trowel can't reach.

Every do-it-yourself cement mason needs other tools, such as hammer, level, measuring tape, wire brush, stringline and the like. You more than likely already have these.

Clean your tools right after finishing a job. A good method is to put everything in the wheelbarrow and fill it with water. Clean all the hand tools first. If stubborn concrete deposits cling, wire-brush them off. Finally, clean the wheelbarrow. Put tools in the sun where they'll dry quickly.

If steel tools won't be used again for some time, store them in a dry place. Otherwise oil the metal lightly to keep it from rusting.

A wood float, combined with a steel trowel, is used to put a dense finish on concrete.

A metal float imparts a nonskid texture to concrete; is used on paths and sidewalks.

Having two steel trowels makes job easier. Start with 18″ x 4½″, end with 12″ x 3″ size.

Exterior plywood kneeboards are a must to allow you to work on large concrete slabs.

Grind off pointed trowel to make a round-nosed trowel; a hand tool for clean-up.

PLACING, FINISHING AND CURING

To make a lasting concrete slab planning and timing are essential

One rule to remember in finishing concrete: work the surface as little as possible to get the finish you desire. Sounds simple and is. But the temptation to overwork concrete is tough to put down. Don't let it hang you up.

Your choice of finish should depend on your skill with a trowel. The very hardest finish to make is the steel-troweled smooth surface you see on most concrete. This finish is fine indoors for basement floors, garages and such. It is hard, takes lots of wear and cleans easily. But it's too slippery for use outdoors. For patios, sidewalks and driveways use a nonslip finish. One of the easiest nonslip surfaces to make is the metal float finish. That's METAL float, not steel trowel. Use an aluminum or magnesium float. A broomed finish is easy to make, too.

You can use a nonslip finish inside and avoid the hard-to-make steel-troweled finish completely. The surface will be harder to sweep and wash down and not as comfortable for rollerskating. Otherwise it will be serviceable.

If you must have the smooth-floor look, it might pay to hire a skilled cement finisher by the hour to produce it for you. You CAN do it yourself, but don't try to do too much in one day. An area of more than 100 square feet can get away from you on a hot, windy day. The part you gave a first-troweling may get past the stage for its second troweling before you finish the first-troweling elsewhere. This can happen to pros too. Watch it.

If you plan a float or broom finish, you can handle much more without danger of "losing it." Cool, damp weather helps too.

SLABS

Most concrete projects you'll build will be slab work. Every slab should be cast on a firm, dry subgrade (in-place soil) or subbase (special material). If your soil is sandy and well drained, you can cast a slab right on the ground. First remove all sod and vegetation. You can also cast on the ground if you live in a nonfreezing climate. Clay and other poorly drained soils call for a compacted 2-inch layer of crushed stones, gravel or sand to keep water away from the bottom of the slab. Sand is easier to shovel and grade but it should not be very wet when concrete-placing begins. Just damp.

"Mucky" soil should be dug out and replaced with gravel or crushed stones in 4-inch lifts. Tamp each one well.

Build forms for 4-inch-thick slabs out of 2x4's. For 6-inch slabs use 2x6's. These give slightly less thickness than the full 4 or 6 inches, of course, but that's the name of the game. Don't worry about it.

Brace forms by driving 1x2 or 2x2 wood stakes into the ground every 4 feet. In nailing through the stakes into the form it helps to back up the form with the head of a sledge hammer or other heavy object. Forms will strip easier later if you use double-headed form nails. Oil the forms with old crankcase oil so they'll strip from the concrete without sticking.

Gently curving forms can be made of doubled-up 1-inch lumber in place of the 2-inch. For sharper curves use ¼-or ⅜-inch plywood or hardboard bent and staked to the desired radius. Build up two or more layers until you have enough thickness to take the pressure of fresh concrete. The sharper your curve the less thickness is needed.

Forms for slabs may be level or sloped for drainage. In order to drain well, you'll need a slope of ⅛ to ¼ inch per foot.

TRANSPORTING CONCRETE

If you use ready mix, try to have it dumped directly into the forms. This saves work. Otherwise you'll have to wheel it in and dump from the wheelbarrow. In a pinch you can use a bucket brigade, if you have enough helpers. It's hard on backs and buckets. Mix-your-own concrete will probably have to be transported by wheelbarrow.

Concrete should not be dumped in separate piles and raked together. Nor should it be placed in one pile and pushed or allowed to run into place. This practice makes weak slabs. The less pushing, shoving and raking you do, the better for both you and the concrete. Dump concrete against the forms as near to grade as possible. Start so that each succeeding load can be dumped against the previous one. This tends to compact the mix into place.

The ready-mix truck can be maneuvered to place the concrete just where you want it.

If concrete can't be dumped directly onto subgrade, transport it with a wheelbarrow.

Any raking should be with a shovel or hoe. Never with a rake. Don't let water collect at the ends and corners of forms.

Whatever you do, keep your bare hands and feet out of concrete. Concrete not only abrades your skin, it eats through. That hurts.

Never place concrete on wet or frozen ground. On the other hand, in dry weather, dampen the subgrade slightly to keep it from drawing water out of the mix.

FINISHING

After placing concrete, strike off the surface to the correct elevation. Form tops should be at this elevation so they can be used for supports for the straightedge. Place the strikeboard across the forms and begin see-sawing along them. Advance an inch or so with each stroke. You'll soon see a roll of excess concrete form ahead of the screed. With air-entrained concrete this roll is typically round and plump, telling you there's air in the mix. The roll of extra concrete fills in any low spots. If the roll gets too big, shovel some of it away.

Right after strike-off, get going with your bull float or darby. It works mortar to the surface while putting down ridges left by the strikeboard. The leading edge of the bull float should always be raised. This is done by raising or lowering the handle as it is pulled or pushed.

The secret to handling a darby without messing up your newly struck surface is to use a light touch. Hold up some of the weight of the darby if the mix is wet. Bear down a bit if it's stiff. Don't disturb the elevation, just smooth off the ridges. You'll soon get the hang of it.

EDGING AND JOINTING

Soon after bull floating, run all edges

An asphalt-fiber isolation joint must be used between a new pour and the old slab.
Portland Cement Assn.

For a smooth edge, work the concrete up and down with a shovel next to the forms.

and joints with cutting and rounding tools. This is done by moving each tool back and forth on top of the slab. Always use a jointer against a straightedge as a guide. An edger is run using the form as a guide. You needn't try for a smooth finish on the initial run. If you merely get the big stones out of the way, you've accomplished the purpose.

As you move the tool, keep from pressing hard. The flat portion is supposed to ride over the surface not make a deep impression in it.

Control joints should be placed 10 to 15 feet apart on floor, driveway and patio slabs, 4 to 5 feet apart on sidewalks. If you space them too wide, you'll get intermittent cracks at random throughout the slab. Control joints put these cracks where you want them.

Try to locate joints at weakened planes in the slab. For instance, if there's a planter opening in a patio, joints should be arranged to come at this spot. That's where the slab will want to crack. Help it.

FLOATING

When the jointing is done, take a break. It will be a matter of minutes or hours until the surface is ready for hand-floating, the next step. If your mix was dry, the sun is shining on the slab and the weather hot and dry, you have only a little time. If your mix was wet, the slab is shaded and it's cold and damp, you have a long wait.

The purpose of floating is to compact the surface, remove humps, fill hollows and bring mortar to the surface for subsequent finishing. Swing the float over the surface in overlapping arcs. With air-entrained concrete, a lightweight metal float works best. A wood one may tend to tear the surface. Floating gives a gritty texture to the surface and is great as a final finish for a sidewalk, driveway or patio.

The exact timing of hand-floating is puzzling to most of us do-it-yourselfers. The tip-off is how the surface looks. It's different for air-entrained and nonair-entrained concrete.

Air-entrained concrete—Hand-floating operations should start before the surface gets too dry and tacky. Finishing may begin as soon as the concrete has begun to set. A good guide is to put one of your kneeboards onto the slab and kneel on it. When it sinks only 3/8 inch, start floating. If it sinks more than this, float over the depression to hide it and wait. Make your test in the first-placed section of the slab. This usually begins setting up first. Start your finishing there, too.

Strike the concrete off with a straightedge resting it on the 2 x 4 forms on either side.

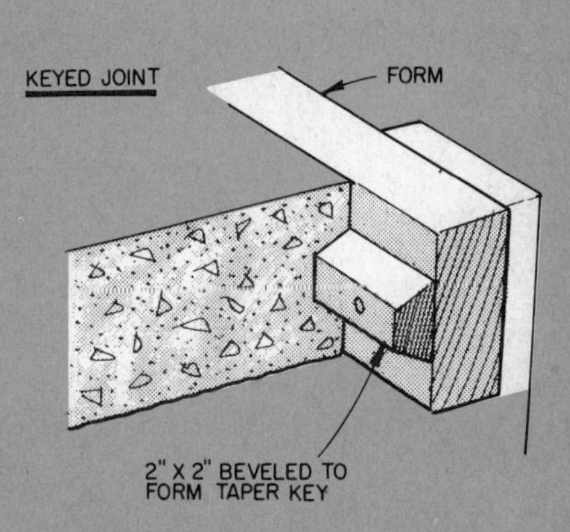

Nonair-entrained concrete—Water bleeding out of nonair-entrained concrete delays finishing. Never float a slab with surface water on it. Wait for the water to evaporate and for the water sheen to disappear. Squeegee it off by dragging a garden hose across, if it persists. The surface should be dull for finishing, not shiny.

STEEL-TROWELING

Where you want a smoother finish, you'll have to follow floating with steel-troweling. Begin troweling only after the concrete has set enough that an excess of cement, water and fine sand isn't brought to the surface. This is normally some time after floating, when the water

Raise bull float handle high when pulling toward you, lower it when pushing away.

Use easy stroke with a darby so slab is smoothed without changing its elevation.

Snap a line on the just-placed slab for each control joint. Get it square with slab edge.

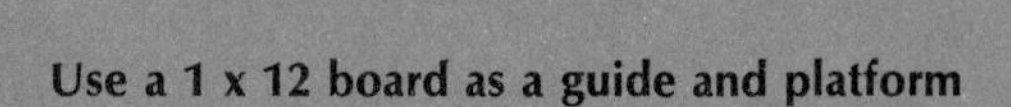

Use a 1 x 12 board as a guide and platform when running control joints with a tool.

sheen brought on by floating has disappeared. Too much "soup" at the surface makes a poor-wearing finish.

During the first troweling, hold the trowel as flat against the concrete as possible without "digging in." If the trowel blade is tilted much at this stage, you may get a "washboard" finish.

When the water sheen disappears again, follow the first troweling with a second. Tilt the trowel more this time because the concrete is much stiffer and can take added pressure on a smaller blade area. Use a smaller trowel if you have one. It's easier on your wrist.

Kneeboards leave marks on the surface. Work backwards, troweling over the marks as you go. Float and trowel

To prevent cracking, control joints should be a minimum of 1/5 the depth of the slab.

Begin hand-floating when water sheen has disappeared. Move float in smooth arcs.

After floating, use a steel trowel on the surface to make a dense, smooth finish.

Kneeboards are moved one at a time. Work backwards to finish over marks left by them.

Float and trowel right over joints, then run them again with a jointing tool. Edges, too.

To keep the excess concrete from hampering finishing operations, clean from the forms.

Lifting the edge of steel trowel on final runs helps to compact the surface concrete.

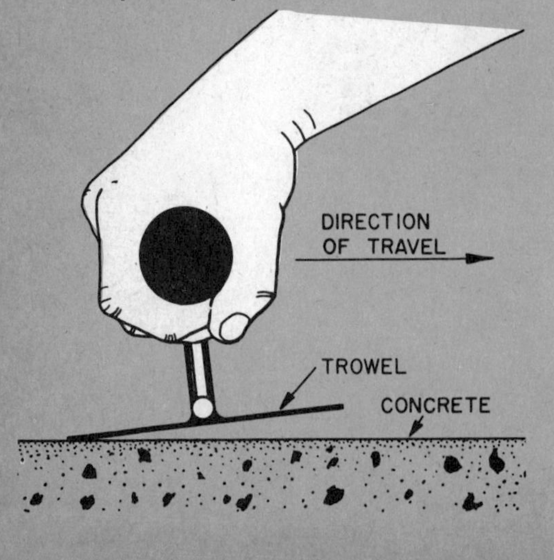

right over your initial grooves and edges, too. Then run them again, leaving everything smooth and finished-looking behind.

You may want more than two trowelings. Allow some time between them to let the concrete gain more strength.

If a really hard finish is needed, retrowel the slab after it has hardened to where no paste collects on the trowel blade. A ringing sound will be made as the trowel is drawn over the surface.

As you trowel, you'll find it handy to support yourself with one hand on a float. The imprint can be troweled out.

When you've placed all the concrete you want for the day, put in a header between your side forms and stop. By tacking a tapered strip of 2x2 along the header, a keyed joint is formed. This joins beautifully with the next slab when you build it.

Every slab that meets an existing slab or wall should be separated from it with what is called an *isolation joint*. You can buy strips of 1/2x4-inch asphalt-impregnated material ready to use. Place them against the existing wall or slab with their tops at or above the level of the new slab. Any excess material can be sliced off after the new slab has hardened. An isolation joint permits independent movement between two concrete structures without restraining them.

SPECIAL FINISHES

You have a choice of an unlimited number of textures and designs for your new slab. Many of them are much easier to make than a steel-troweled finish. They all start after the concrete has been struck off and bull-floated.

Broom Finish—For an excellent broom finish, use a soft hair broom for a mild texture, a stiff bristle broom for a coarse texture. Do the brooming soon after the first hand-floating. A pushbroom works best. Lay it on the surface and draw it across the slab toward you. If small balls of concrete "pill" up from the surface, it's too early for good brooming. Wait. If you have to bear down on the broom to produce a texture, get going. Time is short.

Swirl—Make a swirl design by moving the float, trowel or other finishing tool in overlapping semicircular arcs. Hold it flat on the surface. The smoother the swirling tool, the finer the texture. Swirling is always the final finishing step.

Stipple—This effect is created with a broom. Pound the hardening floated surface with the bristles of a stiff broom. You may soften the effect if desired, by steel-troweling over the stipples.

Rock salt embedment—This creates a texture looking somewhat like rough marble. To make it, scatter rock salt over the fresh floated surface and press it in with a trowel. Don't start the process too early or your salt will sink out of sight into the slab. It should rest on the surface. Later the salt itself washes out, leaving pockets. Don't use this method in freezing climates.

Leaf impressions—Patterns resembling leaf fossils can be made by troweling real leaves into a fresh concrete surface. The leaves should be embedded after the first steel-troweling. Get them deep enough so you can trowel over their tops without dislodging any, but not so deep that mortar is deposited over them. Remove the leaves after the concrete has set.

Incised patterns—Your ingenuity is the only limit to the variety of incised (engraved-in) patterns you can apply to a concrete surface. Joints can be troweled to look like flagstones. Star patterns can be created by jabbing the edge of a trowel into the surface in intersecting lines. Different sizes of cans may be used like cookie cutters to make circle patterns. Kitchen utensils you can pirate and use are tablespoons, zig-zag cutters and forks.

CURING

Curing is one of the most important concrete operations, often the most neglected. The newly placed concrete must be protected from drying out by the sun or by a dry wind. Any method that keeps the concrete wet will work—either by adding water, or by preventing evaporation.

Some common curing methods use burlap or canvas coverings kept continu-

On final troweling, after concrete hardens, tilt handle as you draw it over the surface.

Any marks left can be removed by troweling over them. Support one hand on float.

For a broomed texture, pull a hair broom over just-floated slab in straight lines.

A liquid curing compound may be sprayed, rolled or brushed after finishing operation.

Swirl finish is made by swirling a light weight metal float over the surface. It's a final step.

ously wet for six days. Polyethylene sheeting also can be used, if the edges are held down with piles of dirt or sand. Place the coverings as soon as the surface has hardened enough that it won't be marred. Other methods of curing are coverings of earth, sand or damp straw.

Another method is to place a lawn sprinkler or soaker hose where it will keep the concrete continuously wet for six-day curing period. Leaving the forms in place will cure a concrete wall.

A great new method is to spray, roll or brush on a curing agent. Some are pigmented. Some clear. Pigmented ones go on more evenly. All soon weather away after they've done the job. One good one I know of is W. R. Meadows' Sealtight *Cure-Hard*. Only one coat is normally needed.

Remember, that the longer concrete is cured—even beyond six days—the better it will be. "6667" or "6668" concrete is even better than "6666."

In both hot- and cold-weather concrete construction, special precautions are necessary. You won't feel like working outdoors when the weather is too cold for normal procedures, but you may place concrete during weather that is hot enough to need special treatment. Plan ahead. Have the necessary equipment and materials on hand. Build sunshades, windbreaks. Consider starting the job after the heat of the day has passed. Sprinkle aggregate piles, use ice in place of some of the mix water. Sprinkle the subgrade and forms before placing cement.

Protect against evaporation by covering the surface until you're ready to finish it. Don't delay placement. Don't let the set get ahead of you. Have plenty of help on hand. Start the cure as soon as the surface will take it without damage. All of these steps will make your hot-weather job that much surer of success.

CASTING WALLS

Walls are made differently from slabs. Only the top surface may need to be floated, troweled and edged. The rest of the wall will take on whatever finish is provided by the form. For this reason and others, plywood is an excellent material for forms. It produces a smooth wall surface. What's more, it can be bent

Rock salt, troweled into the concrete dissolves; leaves beautiful textured surface.

Portland Cement Assn.

Flagstone effect is cut into concrete with a bent pipe. Retool after each troweling.

to form curving walls. They're often more pleasing to the eye and are stronger. Order A-C fir plywood and place the "A" side toward the wall. The face grain should run across the 2x4 supports. If you want the form to be reused many times, make it of exterior grade plywood. Plywood ⅝ to ¾-inch thick needs supports every 12 inches. Supports for thicker plywood forms can be placed 16 inches on centers.

A good wall finish free of honeycombs can be produced by tapping the concrete-loaded forms with a hammer. This forces coarse aggregate away from the form and dislodges trapped air. Consolidate concrete in the forms by spading it with a shovel, especially next to the form. For uniformity, place concrete in layers not more than 12 to 18 inches deep along the wall form. Thoroughly consolidate each layer before placing the next one. Do not drop concrete more than three or four feet into forms. Use a chute instead. Dropping makes the coarse aggregate separate from the rest of the mix. You can make the chute out of a 2-inch-thick plank with 1x4's or 1x6's nailed-on the sides.

Portland Cement Assn.

Leaf impressions are made with real leaves. Flatten into surface and finish over them.

USING READY MIX

Order your concrete like the pros do

Any time you're thinking in terms of a cubic yard or more of concrete, you're in the market for ready mix. You can use it in any location accessible to the truck-mixer. You can even wheel it from the truck to your jobsite.

There's a trick to getting quality ready mix. You can't just call a dealer and order so much concrete. That would be like telephoning a pet shop and asking them to send over a pet. You might get a mouse, or you might get a tiger. Ready mix producers these days have sophisticated equipment that can make many different kinds of concrete. You must order the kind you want. Asking for "6666" concrete would be a good start. But there's still a better way.

Order your concrete like the pros do, by specification. To get you the most and best concrete for your money, put a slightly different slant on the "6666" principle explained in earlier chapters. Specify that your ready mix conform to ASTM *Specification C94*. *C94* is familiar to all ready mix producers. They eat it, drink it and go to bed with it. Your dealer will recognize *C94* as the American Society for Testing Materials' *Standard Specifications for Ready Mixed Concrete.* The spec outlines the producer's responsibility for getting quality into the mix. Among the things covered are quality of materials, quality of mix, slump tolerance—slump is a rough measure of concrete's workability—measuring of materials, and much more. The ASTM spec does not cover placement, consolidation or curing of the concrete after delivery to you. These are your hang-ups.

Once you've mentioned ASTM *C94*, tick off these items on your fingers one

Ready mix lets you attend to placing and finishing, frees you from mixing concrete

by one: (1) Six bags of cement per cubic yard. (2) Maximum size of coarse aggregate. This should be one-fifth the cross section of your project. Often 3/4 inch. (3) *Maximum* of six gallons of water per bag of cement. (4) A slump at the point of delivery of 4 inches, plus or minus 1 inch. (5) Entrained air content of 6 percent, plus or minus 1 percent.

Note the difference from mix-it-yourself concrete. These specs are designed to take advantage of the producer's knowledge of his own materials for your benefit. A maximum amount of water is specified. This permits your dealer to use even less water than 6 gallons per bag of cement and get you a stronger mix, provided he can maintain that 4-inch slump (also specified). Too little slump—stiff, hard-to-work concrete. Too much slump—ags settle out, making weak concrete. The tolerances in slump and air content are only fair to your dealer.

Within reason, your ready mix man should be able to deliver concrete to you at the time you and he agree on. If bad weather, equipment breakdowns or other unforeseen problems delay him, he should call you. The obligation works both ways, You should plan to be ready when the truck-mixer arrives. If, for some reason, you can't make it, call and work out a new time.

While you might be tempted to look up a number of ready mix producers in the classified telephone directory and call around for the best price, don't necessarily order from the cheapest one. If you buy ready mix by price alone, you're asking to be taken by one of the few chiselers in the field. Ask for bids only from reputable dealers. Generally the nearer to your house the dealer's yard is located, the more economically he can serve you.

One way to check out your dealer is to ask whether he's a member of the National Ready Mixed Concrete Association. It's a good sign if he is. Talk to

You'll need fewer helpers if the truck can dump ready mix directly onto the subgrade.

Lay out sheets of plywood or planks to protect your driveway from truck's weight.

See that truck has clearance on all sides. Watch for power lines and other obstacles.

When filling a wheelbarrow from a truck, place a plywood board to catch drippings.

READY MIXED CONCRETE

Water added Gal./cu.yd.	Slump increase (inches)	Strength reduced (percent)	psi. reduced
1		4½	150
1½	1	6½	225
2		8½	300
3	2	13	450
4		17½	600
4½	3	19½	675
5		21½	750
6	4	26	900
7½	5	33	1150
9	6	40	1400

CHECKLIST FOR ORDERING READY MIX

1. How much ready mix?
2. When you want it?
3. Where you want it?
4. Conforming to ASTM C94.
5. 6-bag mix.
6. Maximum size of aggregate?
7. 4-inch slump, plus or minus 1 inch.
8. 6 percent entrained air, plus or minus 1 percent.
9. How much will it cost?
10. How you will pay for it?

neighbors who have used ready mix. Who was their supplier? How did they like the service? Was the driver helpful? Don't expect the driver to shovel concrete. His job is handling the truck-mixer. He can, however, help by putting the mix as close to where you need it as possible.

HOW MUCH READY MIX?

Ready mixed concrete, as you probably know, is sold by the cubic yard. You can specify half-yard increments. One problem is that most projects are figured in square feet. A little calculating is needed to tell your dealer how much concrete you need. Measure the area of the job. For slabs measure length and width in feet and multiply to get square feet. For walls measure height and length and multiply to get square feet. Then multiply the square footage by the thickness in inches. Divide this by 300 and you have cubic yards with a 10 percent allowance for waste.

Suppose you build a patio 10 feet by 16 feet and 4 inches thick. Multiplying 10 times 16 gives 160 square feet, the area to cover. Multiplying by 4 (the thickness) gives 640. Dividing 640 by 300 ends you up needing slightly more than 2 cubic yards. Take a chance and order just 2 yards.

When you figure time saved, the cost of ready mix is not much more than were you to buy the materials separately and mix them yourself. Most dealers give five minutes per cubic yard free unloading time. After that you have to pay an extra hourly charge for holding up a truck that should be out delivering another load. If you are more than a minimum number of miles from a ready mix plant, there may be an extra charge for the long haul. Small batches cost more per yard than full truckload batches. Some dealers charge extra for Saturday delivery.

If you can afford it, there's a swinging new way to transport ready mixed concrete. It involves renting a concrete pump. A few ready mix producers now have them. The unit comes with an operator for a $50 or more minimum charge, plus a few dollars extra per cubic yard. Concrete is fed through a long pipe with a flexible hose at the end. You spot the mix precisely where you want it. Concrete can even be pumped up through a window or smaller opening.

GET READY

Plan your job in advance so you'll be ready to go when the truck arrives. If your order will come in two or more trucks, schedule them to arrive about an hour apart so you're not swamped. Hopefully, you'll have a short breather between trucks. Before the truck arrives:

1. Have the forms well braced so they

stay aligned as concrete is being placed in them. Have them oiled and ready for use.

2. Place a wheelbarrow or two handy to transport the mix, if that's necessary. If concrete is to be chuted into place, have a chute ready. Most trucks can reach up to about 12 feet using their own chutes. Some dealers can provide an extra-length chute.

3. Have the subgrade dampened to keep the ground from absorbing water from the mix. See that it's level, or properly sloped.

4. See that your strike-off boards are handy and your screeds are in place. A screed is a temporary 2x4 staked between the forms for a slab. It serves as a support for one end of the strikeoff board when spanning clear across the forms is not practical. After strike-off the screed is removed and a little extra concrete chunked into the space left by it.

5. Have planks and runways in place to protect sidewalks, driveways, sewer lines, lawn, etc., from the heavily loaded truck.

6. Have your curing materials handy to prevent over-fast evaporation of mixing water.

7. Keep a plan in mind as to how you are going to place the concrete. Where first? Where next? Where will you finish up?

8. Have your helpers alerted.

A ready mixed concrete job requiring a truckload or two is not a do-it-by-yourself project. You need help. If the mix can be simply chuted or dumped onto the site, you'll need three men. You'll want to have two more on hand if the concrete must be wheeled in and dumped. If one helper is all you can muster, scale your order back to three cubic yards or less. This is enough work for two men in one day.

If it's a garage slab you're casting, divide it in quarters. Do one quarter each weekend. Put keyed construction joints between the slabs with 3-foot-long No. 4 reinforcing bars across at 2-foot intervals to tie the slabs together. If it's a patio, plan for construction joints to fall where you otherwise would have planned control joints.

You wouldn't believe how fast concrete sets up on a hot day with the sun shining on it. If the temperature is over 85 degrees, do your work in the late afternoon or wait for a cooler day. On the other hand, don't work when the temperature is much below 55 degrees.

Remove debris and unnecessary equipment from the area to give the truck

Good ready mix is too stiff to be poured, so have plenty of help on hand to shovel.

An extension chute lets you reach the job, saves lots of hand wheeling and shoveling.

more freedom of movement. In planning the truck's access, avoid soft, spongy ground. If you order a truck into a bad spot against the driver's advice and it gets stuck, the towing charge is yours.

WANT A WETTER MIX?

Watered down ready mix is about as worthless as watered down booze. You have allowed your ready mix producer as many as six gallons of water per bag of cement. If he can't make good, workable concrete with that, he'd better switch to managing a diaper service. Next time buy from someone else. Just be sure, though, that your idea of "workable" isn't "pourable." Good ready mixed concrete cannot be poured or flowed into place. You have to wheel it or shovel it or rake it.

If you're in doubt about your concrete's slump, make a slump test. To do it, buy or build a slump cone. Just having it on the job will establish you as a real pro. Design it out of 16-gauge galvanized metal according to the drawing. Solder the joint. The ramrod can be made from a 2-foot length of 5/8-inch steel rod by grinding a bullet-shape on one end.

Scoop concrete right off the chute of the truck as it discharges into a wheelbarrow. Place this concrete in the slump cone in three equal amounts (not equal heights. Rod each lift 25 times (count'em) to consolidate it. Strike off the concrete at the top of the cone and lift it gently away. The concrete will sag down as it loses support. Measure the distance from the top of the slump cone down to the top of the pile of concrete. This is your concrete's slump. It should measure between 3 and 5 inches, according to your specs.

If the slump is off spec, you can do one of two things: (1) call the plant and send the load back or (2) add a limited amount of water to bring a too-low slump to where it should be.

If you add water, do it the right way. When ordered according to the directions mentioned previously, chances are that your mix is under the 6-gallons-per-bag maximum. You can, with a clear conscience, add enough water to bring it up to the maximum 6-gallon figure. Don't at one time add more than 1 gallon of water for each cubic yard in the mixer. It's better to have the driver pump in a little water, then add a little more later, if necessary. After adding water, have the driver run the drum at mixing speed for at least 2 minutes before discharging any concrete. It's important to mix the water in thoroughly.

Never drop concrete more than a few feet. Build a chute if necessary, with planks.

As a rough rule of thumb, to increase the slump 1 inch, add a total of 1½ gallons of water for each cubic yard of mix in the drum. Each gallon of water added will lop 150 psi. off the strength of your hardened concrete.

If the mix contains the maximum 6 gallons of water per bag of cement and still is unworkable, you can grit your teeth and add up to 2 gallons per cubic yard more water in two steps. Test the mix for slump in between. For instance a 3-cubic-yard load could have up to 6 gallons of water added if need be. While this will reduce compressive strength of the hardened concrete some 300 psi., it is permissible, say the experts. But no more. Always mix for the minimum 2 minutes after adding water.

HAUL-IT-YOURSELF

If a small amount of ready mix is what you want, there's a new way to fill your need. It's catching on across the nation. Haul-it-yourself ready mix, as the method is called, is an ideal way to buy a cubic yard or so of concrete. You get a trailer-mixer filled at the plant with the ingredients for a good concrete mix. You tote it home behind your car. As you drive, the trailer does the mixing. When you get home, the load is ready to dump. Each trailer has its own mechanism for dumping.

While a trailer will hold from 1/4 to 1 cubic yard, it is light and easy to maneuver.

There are really two types of haul-it-yourself concrete. One uses trailer-mixers with tilting drums. The other uses hopper-type trailers with hydraulic dumping mechanisms. Avoid using hopper-trailered concrete, if you can, because no mixing takes place in the trailer. As you jostle along, coarse aggregate settles to the bottom of the trailer and mortar works to the top. What you have by the time you get home is unmixed concrete. If you must use nonmixing trailers, order a mix that's so stiff the aggregates can't possibly settle. Specify a slump of something like 2 inches maximum. You'll have one heck of a time placing and finishing this mix, but you'll have quality.

You can make your own hopper trailer for ready mix out of any trailer by lining the insides with polyethylene sheeting. Have ready mix placed in the sheeting. Without a dumping mechanism you'll have to shovel the mix out when you get home—a definite drawback.

The cost of haul-it-yourself trailer-mixed concrete is only slightly more than for regular ready mix. You ordinarily get a half-hour of trailer rental included. Additional trailer time is extra. Prices go up a bit on Saturday, when the mixers are especially popular. If you don't clean the mixer thoroughly before returning it,

Truck driver can work discharge lever to give you a little or a lot of ready mix.

Driver can rinse cement from the truck into a wheelbarrow, using self-contained water.

Haul-it-yourself trailer-mixer hooks to your car and backs up to most jobs. It mixes as you drive.

Trailer-mixers are great for small concrete jobs needing from ¼ to two cubic yards of concrete.

Before returning mixer, make a thorough clean-up. Avoid a probable extra charge.

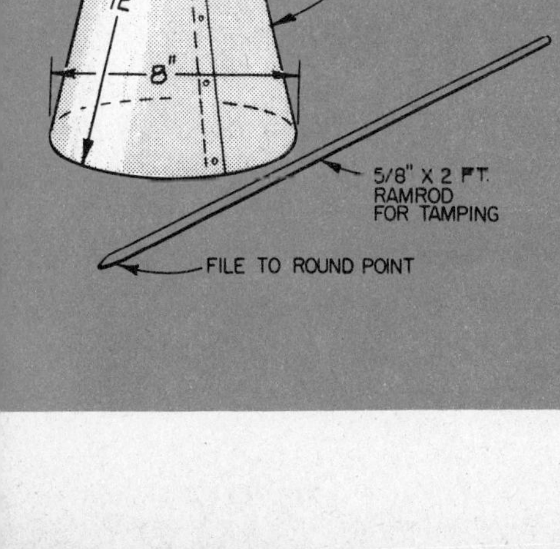

MAKE A SLUMP CONE

MIX SHOULD SLUMP 4"

4"

4"

12

8"

16 GAUGE GALV. SLUMP CONE

5/8" X 2 FT. RAMROD FOR TAMPING

FILE TO ROUND POINT

there will likely be an extra charge for this.

Smaller trailer-mixers let you pace the work giving ample time for proper placing and finishing. You can do an entire job, such as a patio or garage slab, in sections without bugging your friends to help. Often the mix can be dumped between the forms without the use of a wheelbarrow.

Haul-it-yourself concrete is great where a heavy ready mix truck would have to back over a sidewalk or driveway and risk cracking it. The much lighter trailer won't hurt any paving that a car could drive on.

A universal trailer hitch attached to your bumper pulls the trailer. Most cars can be fitted. Though the trailer-mixer weighs more than two tons when fully loaded, a full-sized car can handle it if starts and stops are made gradually.

One caution: Some systems for furnishing haul-it-yourself concrete are being purchased by other than ready mix producers. Be very wary of their ability to sell you quality concrete. After reading the first few chapters of this book, you'll know much more about concrete than many such producers. Ask a few pointed questions about selection of aggregates, what exactly goes into the mix, slump, air entrainment and quality control. Then decide whether to spend or split.

While you can probably handle haul-it-yourself concrete alone, the ideal crew is three men. Two men handle placing and finishing. One man takes charge of getting the trailer, bringing concrete, cleaning the mixer and returning it.

It's a good idea to reserve a trailer-mixer ahead of time if you want to use it on a weekend.

Many ready mix producers will rent you implements to place and finish concrete, things such as a wheelbarrow, a power trowel or a bull float.

Ready mix producers can furnish colored concrete at slight extra cost. Yours probably has samples of the colors available. To find a dealer near you, look in the classified telephone directory under "Concrete, Ready Mixed."

HOW TO AVOID CONCRETE

Surface defects show that steps to quality

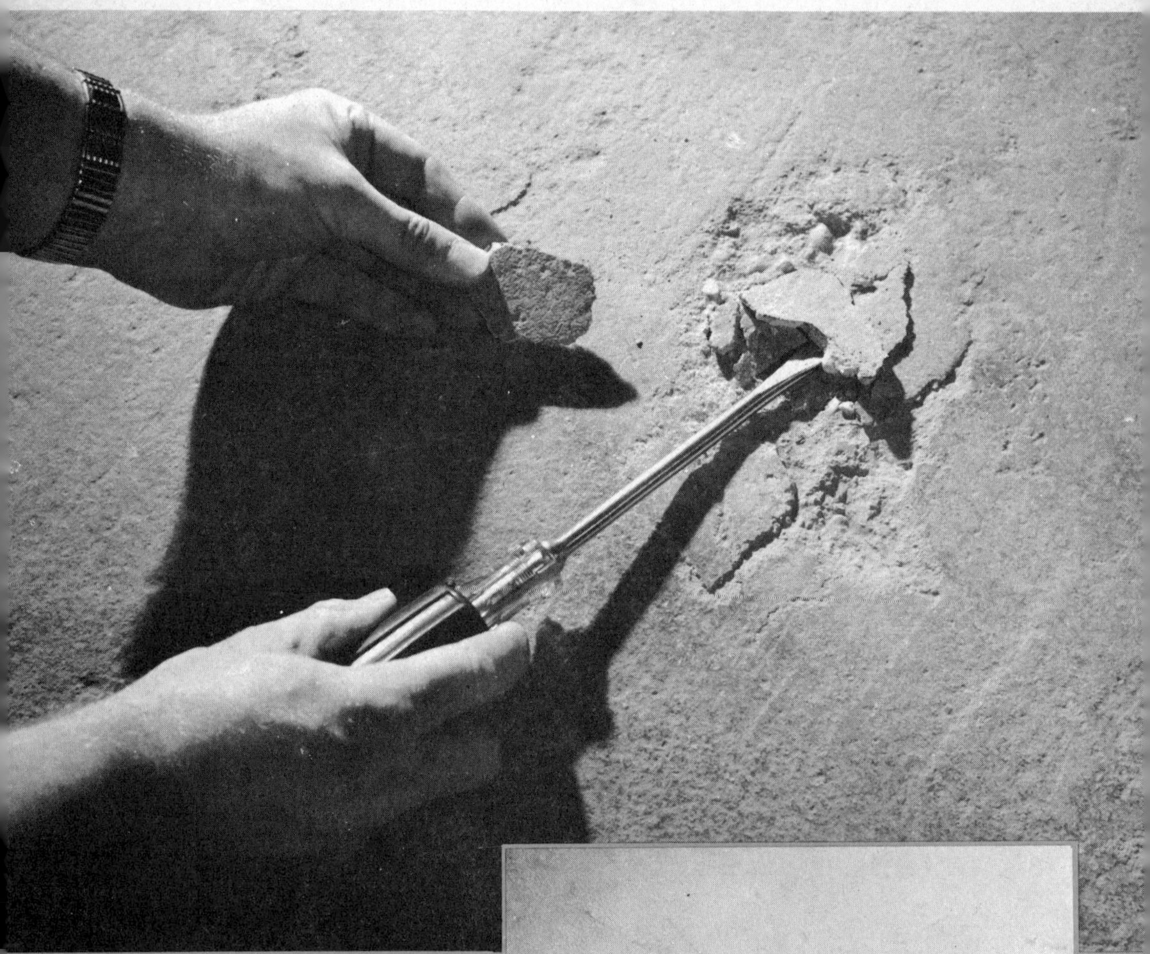

Scaling

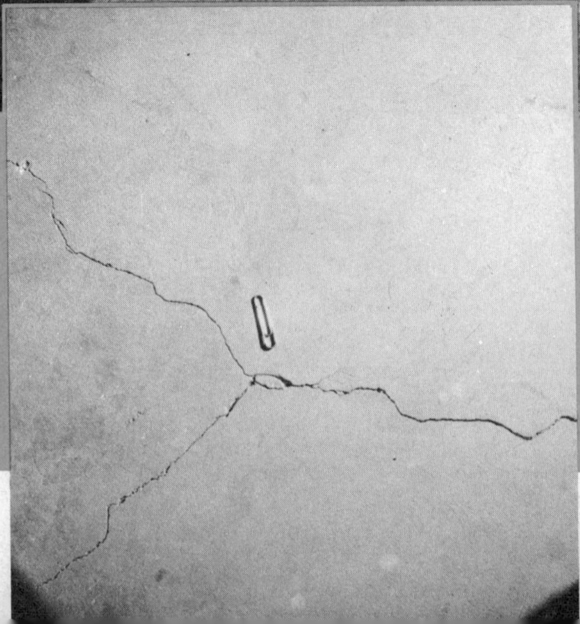

Cracking

FAULTS

construction were ignored

Roughness

Dusting

Faults in concrete can be avoided by the quality approach. Some faults are serious. Others are simply unsightly. They all tell a sad story about how the concrete was made or used. Good concrete does not develop faults.

Here are some of the faults commonly found in concrete slabs and walls. The photographs illustrate them.

Scaling: In scaling, the surface of hardened concrete breaks away from the slab to a depth of about 1/16 to 3/16 inch. It usually happens before the slab is very old and is greatly aggravated by deicers. Keep concrete from freezing during its curing period. Use air-entrained concrete with 6 percent air. Don't start finishing while free water is on the surface. To save an existing slab from scaling, spray on a thin coat of linseed oil thinned with an equal amount of turpentine. (Be careful. This mixture is flammable.)

Roughness: A certain amount of surface roughness may be desirable for slip-resistance. Too much roughness leaves the surface coarse, ragged and hard to clean. It usually comes about when hot, dry air evaporates moisture from a slab, making the concrete set faster than it can be finished. Have enough help on hand to keep ahead of set.

Cracking: Not a surface defect, but one that reaches through the slab, cracking usually comes from weak concrete and improper control joints. Excessive loads, or loads placed on hardening concrete, too early, also will cause cracking. Be sure that the concrete is thick enough for the job it must do. Use "6666" concrete. Provide control joints in slabs (10-foot-square panels should be the maximum).

Dusting: This is where the surface is soft and wears off easily as a powdery material. A dusting concrete surface scratches easily with a knife. It indicates low-strength concrete probably having been made with too much water, receiving improper curing, too much finishing or having silt in the mix. Use clean, well-graded fine and coarse aggregates. Use no more than 6 gallons of water per bag of cement. Don't finish concrete when it has water on the surface.

Structural wall crack

Crazing

Structural wall cracks: Walls as well as slabs need control joints to handle cracks resulting from shrinkage. A small wood insert put vertically in the wall form creates a weakened plane in the wall where it will crack. Later the crack can be sealed with an elastic calking compound. As in slabs, control joint depth should be one-fifth the thickness of the wall.

Crazing: Crazing is when small shrinkage cracks, not deep, form a pattern in the surface. They take on an "alligatored" look, especially when wet. The cracks are caused by rapid surface drying when the slab was built. This comes from hot sun, low humidity or drying winds. They're also caused by working nonair-entrained concrete while the surface is too wet. Overworking with a darby or bull float can bring too much mortar to the surface and permit crazing. Don't start troweling concrete until the water sheen has disappeared from the surface. Use the bull float and darby once over only.

Spalling: Spalling often occurs where a slab meets a concrete wall. The slab shrinks initially. Dirt gets into the open joint. Then when the slab later expands in the heat of the sun, chunks of the slab's top are chipped or spalled off. Use soft, thick isolation joint material at every joint between a new slab and an existing wall or slab.

Popout: Hard to tell from minor scaling, pieces of the concrete surface chip away exposing the stones underneath. Popouts are caused by the use of a coarse aggregate that can't take frost and swells up when frozen. "Pop" goes the concrete above it.

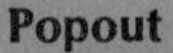

Spalling

Popout

MASONRY TOOLS

A set of good tools is a must for better work in brick and stone

Masonry doesn't require very many tools. Just a few. The same ones serve bricklaying, concrete block-laying and building of stone masonry. Once you have them you're set to go.

Mason's trowel—The most important tool is the trowel. It's different from a cement-finishing trowel, and for a different purpose. In masonry the trowel is used for carrying, spreading and trimming mortar, not primarily for finishing.

The mason's trowel is pointed at the outboard end and available in several sizes. Smaller trowels are easier on your wrist muscles but call for more trips between the mortarboard and the wall. Get the largest trowel that feels comfortable. The size ordinarily used by masons is about 10 inches long from point to heel and 5 inches across the heel. There are two popular patterns of trowel. The "Boston" pattern has a curving blade. The "Philadelphia" pattern has a squared-off blade. If you have much work to do, light weight is a factor in trowel selection. Generally, the lighter trowels cost more.

To use a trowel grasp the handle between your thumb and first finger with your thumb resting on top and well forward toward the end of the handle. Bend your other fingers around the handle—not rigidly—to stabilize the tool in your hand. Practice using the trowel by cutting and working mortar on the mortarboard.

Level—Your wall can only be as good as the level used to build it. Masons use a 4-foot-long wood level to span over a number of units. This aids in aligning them. However, you can use a shorter 2-foot carpenter's level with success. The directions given in this book are designed for use by nonprofessionals and don't require the long wood job.

The level is used to position footing forms, plumb corners, level courses, level individual units and check units for alignment. Its body must be straight.

Mason's trowel is used to handle mortar, tap units into position and to cut bricks.

A 4-foot level is preferable for masonry work but you can get by with a 2-footer.

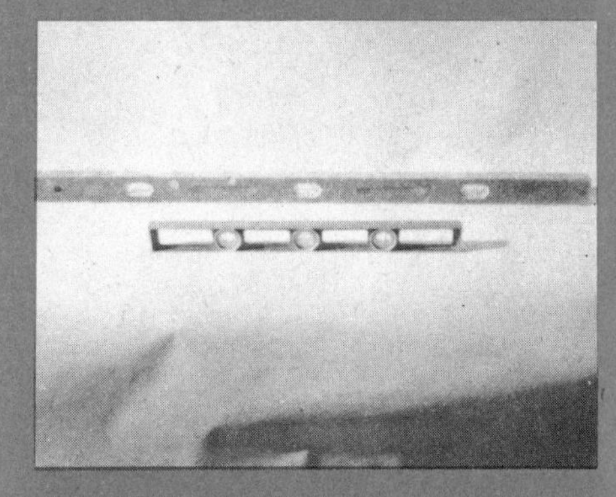

A "spacing rule" is marked with numbered course heights all the way up its length.

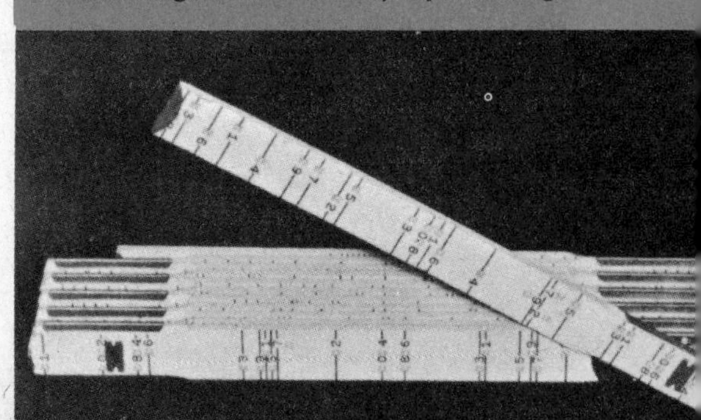

Rule—Whether you use a standard carpenter's rule or a mason's "spacing rule," you must have something to measure with. A spacing rule contains a series of marks that represent course heights in the wall, with as many as 10 different combinations of mortar joint and brick thicknesses. It saves separate figuring of course heights with brickwork. Either a 6-foot wood folding rule or a steel tape may be used. The folding rule is easiest for making vertical measurements.

Mason's hammer—This specialized hammer has a squared head on one end for chipping units to size. On the other end is a chisel-point for scoring units to break them along a line. The hammer usually weighs from 1-1/2 to 2 pounds.

Stringline—A tight stringline without sag in the center is necessary for level courses of masonry. Use a light, strong nylon line and a pair of line blocks to hold it at the corners of your wall. The line blocks make it easier to set up your line for each course.

Jointing tool—To do good looking masonry work, you'll need a jointing tool. The ones for concrete block are long and make either rounded or V-joints. The ones for bricklaying are shorter and curved. They too make rounded or V-joints.

Blocking chisel—In cutting face brick where you need a clean, sharp cut, make the cut with a blocking chisel. Its 3- to 4-inch blade width spans across the face edge of the brick. The chisel is struck with a mason's hammer.

Besides these hand tools, you may have use for some sort of scaffold. This is best rented when the need arises.

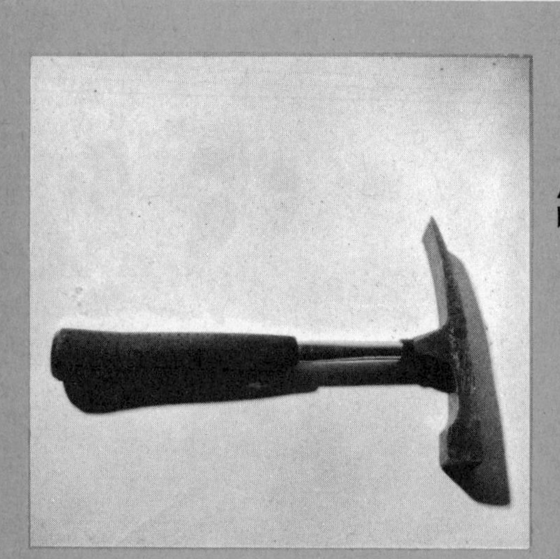

A mason's hammer is used to cut bricks, blocks, stones, to desired size and shape.

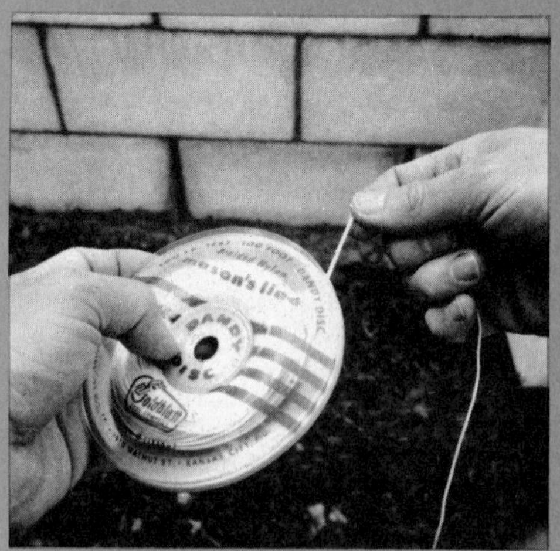

Mason's stringline stored in this plastic reel stays clean and dry when not in use.

Some of the handiest scaffolds are made by laying planks across scaffold brackets hooked to a pair of extension ladders. Always use planks that are free of large knots and other defects. Test them with twice the load they'll be subjected to, mark the top sides and always use them that side up. Never step onto an untested plank.

For utmost safety a scaffold should have a guard rail at the back to keep you from stepping off. Planks should not overlap their supports by more than 6 inches, but they'd better overlap enough to hold. Maximum span for 2x12 planks is 12 feet. Less for lesser widths. Never use brackets nailed on the wall to support you. Likewise, don't use a lean-to or shored scaffold supported by angling boards. These three types of scaffold are not safe. Be careful when working on a scaffold. It's easy to get badly hurt.

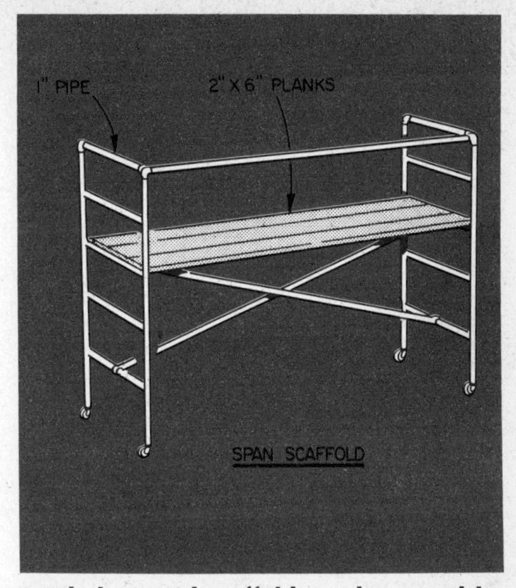

A tubular metal scaffold can be rented by the week or month if work is done up high.

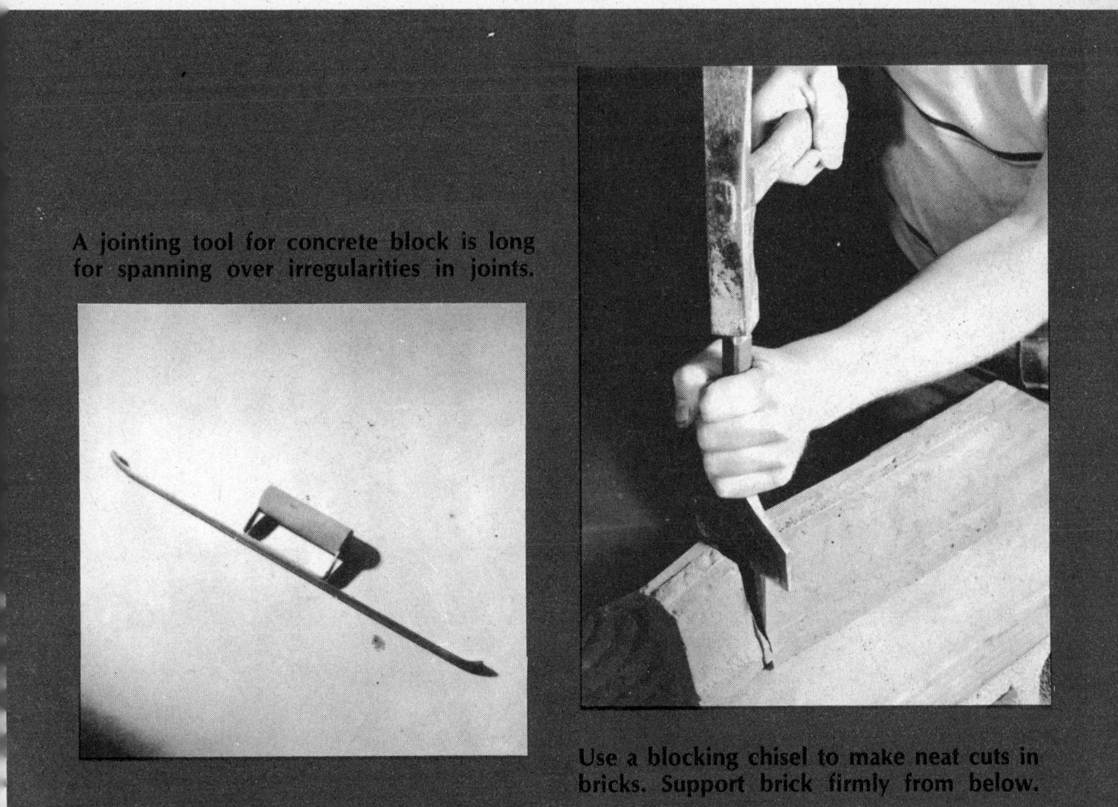

A jointing tool for concrete block is long for spanning over irregularities in joints.

Use a blocking chisel to make neat cuts in bricks. Support brick firmly from below.

LAY BRICKS THE EASY WAY

You, too, can build a square corner and a straight course

At right, once you have the corners built you can lay bricks to a stringline, fast.

Below, by laying bricks with gaps between them, shapely grille walls can be built.

Most people think there's something magic about bricklaying. There isn't. It takes more patience than anything. After a bit of practice, you can do a good job. Professional bricklayers work at a tremendous pace. They can lay more than 1000 bricks in a day. You can't hope to equal that, but there's no reason you can't do as attractive a job at a slower pace.

Assemble the right tools before you start a project. The same ones also serve for laying concrete blocks and stone.

Always lay bricks in full mortar *bed joints* (the ones below each brick and full mortar *head joints* (the ones between the ends of bricks). Mix the mortar according to directions on page 17 about mortar. Lay bricks on a solid concrete footing placed below the frost line or at least 12 inches deep. Come up to grade with concrete, concrete blocks or bricks that can take freezing and thawing in contact with the ground. Many bricks cannot.

Begin by laying the corners of your brick wall. Try, if you can, to space them an even number of bricks and head joints apart. An even number and a half is often okay too. The width of the mortar end joints can be varied to stretch or shrink the measurements enough to avoid piecing. Bricks can have joints from ¼ to ½ inch and more. The wider the joint, the more mortar-like and the less brick-like the wall will be. Wide joints help to cover up irregularities in both the bricks and the bricklaying. Aim for ½-inch joints, lacking another goal.

Brick textures come in all kinds, from a smooth to rough. See samples at dealer's.

Spread mortar on the footing and set one corner brick. Get it right on location. Tap it down to the right height. Check by measuring. If the footing is level, you can measure up from it to the top of the brick. Level the brick lengthwise and crosswise. Butter the ends of three other bricks with mortar one by one and lay them against each other. Slide them together until the head joints close to the proper spacing. You'll soon be able to judge this well enough by eye, but you may want to measure the first joints to be sure.

Then place your level on top of the row and tap the trowel handle gentle on the level until the row settles level with the corner brick.

Also, see that the row of bricks is aimed straight for the opposite corner of the wall. Check by stretching a stringline or snapping a chalk line on the footing.

QUANTITIES NONMODULAR BRICK— 2¼x3¾x8 INCHES

Joint Thickness	Brick Per Sq. Ft. of Wall	Cubic Feet of Mortar per 100 Sq. Ft. of Wall
¼	7.7	4.4
⅜	7.2	6.3
½	6.8	7.9

Now lay four bricks from the same corner in the other direction. Use a carpenter's square to get the corner started right. Tap the row of bricks down level with the corner brick. Line them up aimed straight for the opposite corner.

Spread mortar on top of the bricks at the corner just placed and lay a second-course corner brick on it. Establish the height with a rule. Ordinarily this corner brick should lie in the opposite direction from the corner brick below it. The effect is to lock the corner together like a dovetail joint in a well built drawer. See that the growing corner starts off plumb on both sides.

After laying the corner brick, lay the others in that row as far as you can go without having a brick stick out beyond those in the course below. Level them with the corner brick but be careful not to disturb it. See that all bricks are in line, using the level as a straightedge. See, too, that they're directly above the bricks below by using the level as a plumb. Tap with the trowel to move any.

Keep going, laying bricks both ways from the corner until you end up with a pyramid of four or five courses of bricks topped by one single brick. That's as far as you can go on this corner. Do likewise at the other corners.

Before the mortar sets, check the wall vertically, horizontally and at an angle to see that all the bricks are aligned, level and plumb. If they're not, adjust any of the offending bricks by tapping them. Don't move a brick after the mortar has set. It breaks the bond. Instead, yank the out-of-place brick and those above it and re-lay them.

If the wall is more than one wythe (tier) thick, build both wythes up at the same time. Their head joints should not line up, that is, they should be staggered between the two wythes. The courses, though, should be level with each other all the way up the wall.

When you get the corners built up, you can begin laying in-between bricks. These go quickly. The leveling and plumbing is taken care of by a stringline stretched between the corner bricks. Position it flush with the tops of the cor-

Start a new course by putting down enough mortar for one brick to cross joint below.

Press corner brick down in mortar to proper height on a story pole or brick rule.

Take great care in getting every corner brick level and exactly as it should be.

When corner brick is placed and leveled in two directions, plumb it from two sides.

Build up three or four bricks from corner and tap them down level with corner brick.

Use level as a straightedge to align row of bricks. It's best to use a long level.

Make sure tail end of the lead is plumb with bricks below. Adjust if necessary.

Dump enough mortar on wall to lay four or five bricks to a line. Don't furrow mortar.

Butter mortar onto end of each brick as you lay it. Mortar should be workable.

When lowering the buttered brick toward bedding, push toward the just-laid brick.

ner bricks. Lay the new bricks in mortar and tap them down until their tops are even with the stringline. Leveling each brick across is best done by eye. You may want to check yourself with a level occasionally. Always use a stringline to lay in-between brick. Never trust your eye to get straight rows. Don't set bricks so they crowd the line. Leave a 1/16 inch gap.

QUANTITIES MODULAR BRICK WITHOUT HEADERS

Nominal Brick Size Hgt. Th. Lgth.	Brick per Sq. Ft. of Wall	Cubic Feet of Mortar Per 100 Sq. Ft. of Wall		
		Joint Thickness		
		¼ in.	⅜ in.	½ in.
2⅔ x 4 x 8	7.4	4.6	6.6	8.3
3⅕ x 4 x 8	6.2	4.0	5.8	7.3
4 x 4 x 8	5.0	—	5.0	6.3
5⅓ x 4 x 8	3.7	—	4.1	5.2
2 x 4 x 8	6.6	—	7.7	9.8
2⅔ x 4 x 12	5.0	4.2	6.1	7.8
3 x 4 x 12	4.4	—	5.5	7.0
3⅕ x 4 x 12	4.1	3.7	5.2	6.7
4 x 4 x 12	3.3	3.1	4.4	5.6
5⅓ x 4 x 12	2.5	—	3.6	4.6
2⅔ x 6 x 12	5.0	—	9.4	12.2
3⅕ x 6 x 12	4.1	—	8.3	10.6
4 x 6 x 12	3.3	—	6.9	8.9

The last brick laid in a course is called the *closure brick*. To lay it, butter both ends and the two bricks flanking it. Then lower the closure brick, without knocking off mortar, and tap it into position.

As you complete a course of in-betweeners, move the line up one course. If you happen to tap a brick below the line, pull it out, scrape up the mortar, remortar it and lay it again. If you'll do this simple task, you can lay a wall that friends will think is professionally built.

Soon as the whole wall reaches the highest corner bricks, switch to laying corners again. Keep building up corners and laying bricks in-between them until the wall reaches full height.

Get mortar thick enough so that some will squeeze out as each brick is shoved into place. There should be no hollow spots, gaps or pockets in the mortar joint.

Mortar for head joints is buttered onto each brick just before it is laid. Like bed joints, head joints should be full so that some mortar is squeezed out the top as each brick is laid, proving that the joint is full of mortar.

Mortar that squeezes out of the joints is cut off with the trowel, as though shav-

Set brick in full mortar bedding by tapping into position flush with stringline.

After laying brick, slice off mortar oozing out of joint. Return to mortarboard.

ing it with a razor. This mortar may be returned to the mortarboard and reused. Try to trim excess mortar without smearing any on the face of the wall where it later has to be cleaned off.

Mortar joints must be tooled before the mortar sets too hard. "Thumbprint" hard is about right. On a hot, dry day this condition comes quickly. Keep watch on the first joints laid, lest they get too hard before tooling.

It is sometimes necessary to cut a brick. Some bricks can be broken to a rough size and shape by hacking with the trowel blade while holding in the hand. Stronger ones are better and more accurately broken with a mason's chisel and hammer. Or a mason's hammer may be used to score a line across a brick, using the chisel end, until the brick breaks. Sharp quick blows with the head of the mason's hammer will knock off small pieces of a brick that didn't break clean.

MODULAR SIZES OF BRICK
(in the wall)

Brick Type	Thickness	Height	Length
Modular	4″	2⅔″	8″
Engineer	4″	3⅕″	8″
Economy	4″	4″	8″
Double	4″	5⅓″	8″
Roman	4″	2″	12″
Norman	4″	2⅔″	12″
Norwegian	4″	3⅕″	12″
King Norman	4″	4″	12″
Triple	4″	5⅓″	12″
SCR Brick	6″	2⅔″	12″

MORTAR JOINTS

The treatment of mortar joints in the face of a wall affects both the pattern and wall texture. Mortar joint finishes are of two types: troweled and tooled. In troweled joints, the excess mortar is merely cut off (struck) with a trowel and finished with the trowel. For the tooled joint, a tool is used to compress and shape the mortar in the joint. Concave, V and raked are tooled joints. Weathered, flush and struck are troweled joints. A weeping joint is neither tooled nor troweled. (See the drawing of various mortar joint treatments.)

Most bricks need to be wetted before laying them. Unless they are, high-suc-

To lay closure brick, butter all four ends and lower brick carefully into the opening.

Having completed a course of in-between bricks, move stringline for next course.

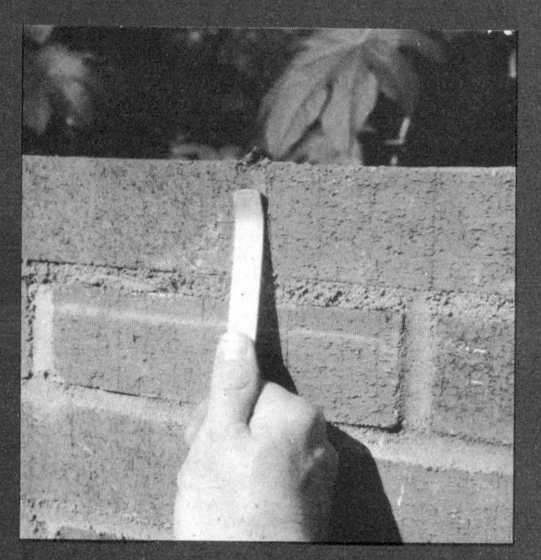

Tool the joints either with concave or V-shaped tool after mortar hardens slightly.

Brush drying mortar on brick faces with soft brush to remove burrs after tooling.

To remove alkaline residue, wash walls after a week with dilute muriatic acid.

Two-tiered walls can be built quicker by backing up bricks with concrete blocks.

tion bricks will absorb moisture from the mortar before its bond has developed. Sprinkling isn't enough. A stream from a hose should be played on the pile until water runs from all sides. After this the bricks should be allowed to surface-dry before they are laid. Water on the surface of the bricks will cause floating on the mortar bed.

A rough but effective test for telling whether your bricks will need wetting is this. Draw a 1-inch circle on the surface of the unit that will be in contact with the mortar. You can use a quarter as a guide. Then with a medicine dropper place 20 drops of water inside the circle. Note the time required for all the water to be absorbed. If it takes more than 1½ minutes, the unit need not be wetted; if less than 1½ minutes, wetting is recommended.

CLEAN UP

Newly laid brick walls need a final clean-up before they can be called complete. The first step is to brush off loose mortar from the bricks before it sets. A bench brush will usually work. If it won't, have a wire brush handy and remove stubborn deposits with it.

Later, after at least 7 days, all mortar stains may be removed by wetting the wall thoroughly, then scrubbing it down with muriatic acid. Add 1 part acid to no less than 9 parts water, then apply the acid to the wall with a household scrub brush. Start at the top and work down. Wear rubber gloves, goggles and old clothes. When the acid has finished reacting, flush it off thoroughly, leaving no acid on the wall. Buff or gray brick should be cleaned without an acid treatment. Try scrubbing with soap and warm water, or consult the manufacturer for cleaning instructions.

The word *bond* when used in reference to masonry, may have three meanings: (1) *structural bond,* the method by which individual units are interlocked or tied together to make the entire wall assembly act as a single structural unit; (2) *pattern bond,* the pattern formed by the units and mortar on the face of the wall; and (3) *mortar bond,* the adhesion of the mortar joint to the masonry units.

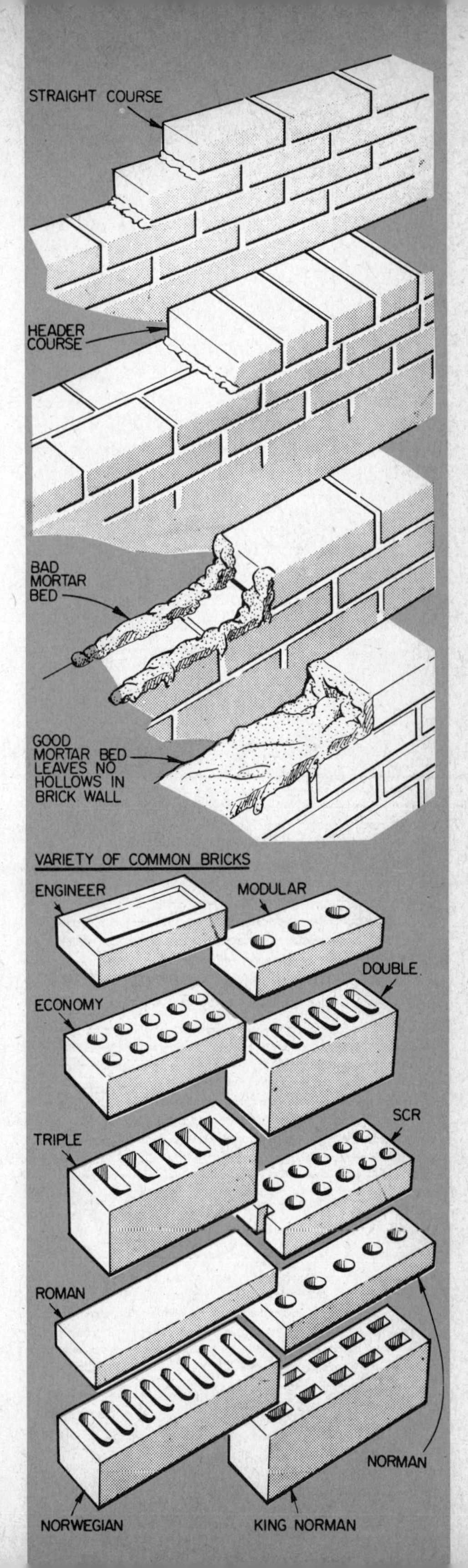

Walls that are two or more wythes thick need something to tie the wythes together. This can be rows of *header* bricks placed in every so many courses. Headers are bricks running crosswise of the wall. Their ends are exposed outside the wall. Modern building codes call for header-bonded walls in which no less than 4 percent of the wall surface is composed of full-length headers. The distance between them is not to exceed 24 inches horizontally or vertically. If you want to avoid the appearance of headers in the wall, the wythes can be tied with metal ties or steel reinforcement. Most codes call for 3/16-inch-diameter steel ties (or other metal ties of equivalent stiffness and strength) spaced so that there is at least one tie for every 4½ square feet of wall surface. The maximum vertical distance between ties should not be more than 36 inches. Ties in alternate courses must be staggered, say the codes.

Check with your building inspector on what you can use. Usually simple walls less than four feet high don't come under the codes. Find out before you build.

Frequently, structural bonds are used to create patterns in the face of the wall. There are five basic structural bonds in common use today. These create typical patterns. They are: running bond, common or American bond, Flemish bond, English bond and block or stack bond. Through the use of these bonds and variations in color and texture of brick and of joint types and color, an almost unlimited number of patterns can be produced.

Most walls have at least two wythes of brick. But under certain conditions a wall with only one wythe will serve. This would be a low garden wall or a higher one laid in a curving or serpentine line to brace it against lateral forces. A wall with built-in columns or pilasters

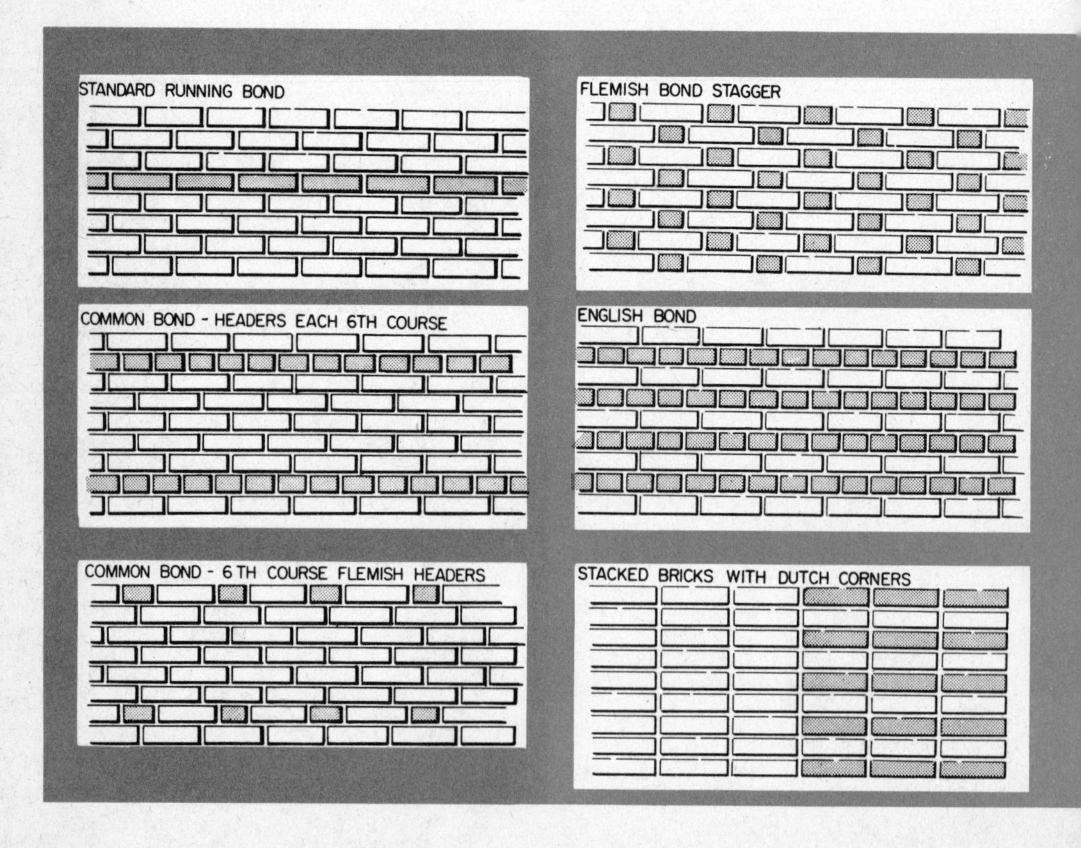

offers more resistance to push-over than one without them.

The Structual Clay Products Institute has developed a special brick that is suitable for one-wythe residential construction. Specially shaped to improve the structural qualities of the wall, it is called SCR brick. One wythe of SCR brick makes a 6-inch-thick wall.

HOW MANY BRICKS?

Quantities of bricks and mortar to order for a project are figured easiest when they are based on the wall area. Modular bricks are easier to figure than nonmodular bricks because the number per square foot is the same no matter what the mortar joint thickness. There are only three standard modular joint thicknesses: ¼, ⅜ and ½ inch. First find the square feet of wall area (length times width, less the area of any openings). Then multiply by the number of bricks per square foot for the type of brick you have selected. Use the tables accompanying this chapter. Do this for both wythes of brick in an 8-inch brick wall.

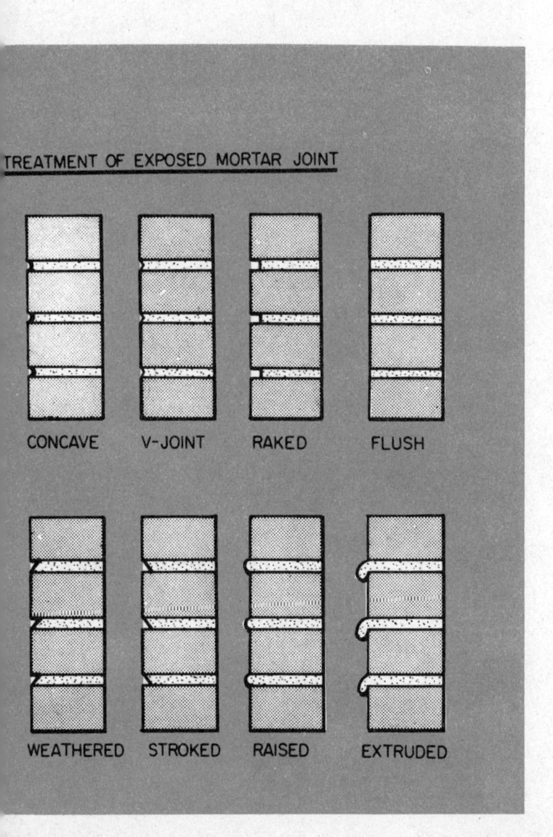

The tables of brick and mortar quantities allow 5 percent for brick waste and breakage and 10 percent for mortar waste. These are good averages. In estimating mortar materials, an approximate rule of thumb is one cubic foot of mortar requires one cubic foot of damp, loose sand. There is about one third as much masonry cement as sand in the mix.

Building brick is made of clay. Clay is one of the oldest building materials, dating back to some 2000 years B.C. Bricks more than 4000 years old have been found still in good condition. Firing in a kiln at high temperatures gives brick its hardness and color. The hotter the temperature, the more weather-resistant the brick. Bricks make good walls and are excellent for paving, too. Their small size lends them to interesting curves, textures and patterns.

There can be lots of variety with bricks. Bricks can be laid into flat, solid walls, with units both protruding and recessed for shadowy textures; they can be laid in curving, serpentine walls; or they can be laid into screen walls with openings for light and breeze. There are smooth bricks, rough bricks, soft bricks, tough bricks. There are glazed bricks so smooth they look like ceramic tile. There are long skinny bricks, short fat ones; big ones and little ones. Just visit a brick dealer and you'll be amazed by the sizes, shapes, textures and colors of bricks.

TYPES AND SIZES

Bricks are covered by national specifications in three grades, based on their resistance to weather, principally frost action. Thesc are *SW* (severe weathering), *MW* (medium weathering) and *NW* (no weathering). *SW* brick should be used for any masonry in contact with the ground. Foundations, patios and retaining walls need *SW* brick. Either *SW* or *MW* brick may be used in walls exposed to the weather above ground, such as house walls, garden walls, barbecues.

Set hard-weathering bricks in mortar on a concrete base for attractive home paving.

A brick swimming pool using reinforced masonry construction needs no finishing.
Structural Clay Products Inst.

Extruded mortar joints takes great skill to make. Avoid them, especially outdoors.

NW brick should be used only where they are not exposed to the weather. Face bricks are made in *SW* and *MW*(but not *NW*)

Bricks should not be laid in freezing weather unless you are prepared to keep the mortar from freezing until it has set. This is best done by covering or heating, or both.

BRICK VENEER

Brick laid over a frame building is called brick veneer. It's a very satisfying do-it-yourself project. If you lay the brick yourself, the cost can be quite reasonable. A brick veneer house combines the weather-tightness and speed of erection of frame, plus the beauty and low maintenance of brick.

A brick veneer structure needs a wider foundation than a frame or solid brick building. This is because an extra width of foundation is necessary for the tier of brick to rest on. While it is possible to cast a separate foundation for the brick or to bolt heavy angle irons to an existing foundation, these methods require a professional's knowledge. Don't attempt them yourself. For a brick veneer wall, you'll need about 4½ inches of foundation extending beyond the sheathing on the outside wall. One inch is for an air space, the rest for bricks.

Since a brick veneer wall is only one wythe thick, it needs no headers. However, it must be tied to the frame wall. For this, 22-gauge corrugated metal ties are nailed to the wall and bent down into mortar joints as the wall is laid. There should be a tie for every 2 square feet of wall area. This means that the distance between adjacent metal ties shouldn't be more than 24 inches, either horizontally or vertically.

Brick veneer needs help from other materials in weather-tightening a house. Before brick veneer is laid, a layer of building paper should be stapled to the wall sheathing. After bricklaying, brick veneer moldings are nailed around openings and calked to seal off the air space. The top of the wall is trimmed with wood and calked. It can really improve a house for not too much money.

HOW TO LAY CONCRETE BLOCKS

Blocks combine beauty with utility, offer imaginative arrangements

Laying concrete blocks is a very satisfying do-it-yourself project. There's nothing difficult about it. And the blocks build so quickly into a wall that you get a real feeling of accomplishment. If you've never tackled block-laying before, you needn't be afraid to try. You can do it.

Blocks can be used for building grille, walls, retaining walls, chimneys, barbecue grills, fireplaces and other such projects. If you have the guts for a big project, you can do a creditable job of building a block basement, garage, swimming pool or even a whole house. Glamor blocks can be used as veneer on the outside of a frame house to increase its value and lower its maintenance. Other blocks make fences, planters or serve as garden edging.

To lay concrete blocks you'll need a minimum of tools. But to lay them without the proper tools would be a disaster. Many of the tools used in laying blocks are the same ones used in bricklaying.

Concrete slump blocks look like adobe brick. Wide joints emphasize texture.

Grille blocks lay into a wall that gives privacy and shade, yet allows ventilation.

Fences of low-maintenance concrete block are becoming popular all over the nation.

The mason's trowel, stringline and level are identical. For block work you'll need a mason's hammer for cutting blocks and a special jointing tool for smoothing the joints. A block jointer differs from a bricklayer's jointer in that it's longer, to span over irregularities in the blocks' horizontal joints. It should be at least 22 inches long and be bent at one or both ends to keep it from gouging the mortar. For tooling concave joints the tool is made from 5/8-inch round bar or pipe. For tooling V-joints the tool is made from ½-inch square bar.

For block work you'll also need a mason's hammer. It has a squared hammer head on one end for chipping units to size. A chisel point on the other end is designed for chipping units along a line to break them.

JOB LAYOUT

Make a rough sketch of every masonry job before you begin. It needn't be fancy. Just so you understand it. The sketch lets you plan your job to suit laid-up dimensions of the blocks. The idea is to avoid cutting units wherever possible. With a careful plan, and the use of full blocks and half blocks, a whole structure can be built without cutting one unit. A no-cut job is a better looking job. Hours of cutting and fitting pieces are saved. The key is planning.

Plan the dimensions of your project with these three rules on mind: (1) An even number of feet, such as 4 feet, 6 feet. (2) An even number of feet plus 8 inches, as 4 feet 8 inches, 6 feet 8 inches. (3) An odd number of feet plus 4 inches, such as 3 feet 4 inches, 5 feet 4 inches.

These rules of thumb apply to vertical measurements as well as horizontal. Remember them.

Never lay concrete blocks without first building a footing of concrete. Cast the footing on firm ground below the frost line. Use quality concrete. In no-frost climates place the bottom of the

Portland Cement Assn.

Homeowner jacked up his house and put a basement under it with 10-inch blocks.

Lay first course of blocks on the footing without mortar, using a ⅜-inch spacer.

Then lay the same blocks in a mortar bedding, furrowing the mortar with a trowel.

Build corners one or two courses ahead. Level by tapping with a mason's hammer.

footing at least 12 inches below grade.

Make the footing twice as wide as the wall. If you're building an 8-inch wall, place it on a 16-inch-wide footing. The footing's thickness should be the same as the wall. Thus an 8-inch wall gets an 8-inch-thick footing.

There's no need to put a troweled finish on the footing. Simply strike it off and float it once. This surface ensures a good bond with the mortar. If there will be a sidewise force against the wall, the footing should be keyed to hold the wall to it. A retaining wall is an example of this type of structure. Make the "key" by embedding a 1×4 board along the footing. Later pull it out, leaving a depression running the length of the footing.

Vertical reinforcing steel, if used in a wall, should start in the footing. When the reinforced cores are grouted, the steel ties the wall and footing together

Take extra care in setting the corners. Plumb corner block from two directions.

Check height of corner block with a rule or story pole with marks 8 inches apart.

Place ribbons of mortar on face shells for laying two or three blocks at a time.

Butter block ends. Mortar sticks best if plunked down and wiped across corner.

into one solid structure. Bend a 2-foot "L" in one end of each rebar. Embed that in the fresh footing concrete midway between top and bottom. The rebars should line up with core holes in the blocks.

Take the time necessary to get your footing level. It's much easier than trying later to level the first row of blocks. What's more, a level footing gives you a level reference point for building the whole wall.

HOW YOU LAY BLOCKS

Start block-laying by making a dry layout of blocks without mortar on the finished footing. Space all joints just as you want them, usually ⅜ inch. Most concrete blocks are designed for use with ⅜-inch joints. They come out in even dimensions that way. The dry layout lets you make small adjustments to get equal mortar joints without cutting any blocks.

Lay intermediate blocks to a stringline stretching between the two corner blocks.

Shove each block sideways against last-laid block; rock gently toward the line.

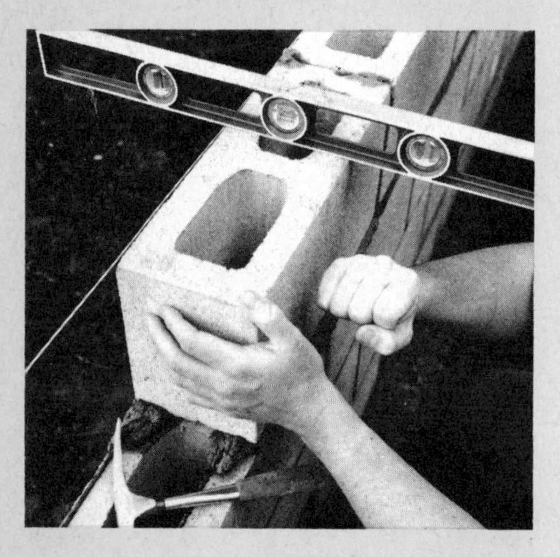

Pounding with your fist makes minor side adjustments. Position each block to line.

To lay block in course, butter all edges involved and lower block into place.

When the joint mortar is thumbprint hard tool the mortar to smooth and compact it.

Use curved end of jointing tool to smooth the short vertical joints. Do them first.

Top side of blocks have wider shells, at right, and are smoother than bottom, left.

Cap blocks give a finished appearance to the wall. Lay them like regular blocks.

When you're sure of the positioning of each block, spot it carefully by chalk-marking before removing it from the footing.

Mix your mortar according to the directions detailed on page 17. Mortar for block-laying should be somewhat stiffer than for bricklaying. A concrete block is heavier than a brick and puts more pressure on its mortar bedding. If the mortar is too wet, it'll squoosh out of the joints. You'll have one heck of a time holding your joints to their 3/8-inch thickness. They'll want to settle to ¼ inch and less. If the mortar is too stiff, you'll have to pound every block into shreds to close the joint to 3/8 inch. Once you get the consistency of the mortar right, try to standardize your proportioning to keep it that way.

Blocks laid directly on a footing get what is called *full mortar bedding*. Blocks laid on top of other blocks get *face shell bedding*. A few codes still call for full mortar bedding on all courses. Better check yours. A course is one row of blocks, horizontally.

For full mortar bedding spread the mortar out to the full wall thickness. Every portion of the block's lower shell rests in mortar. For face shell bedding spread the mortar only along the two outer edges of the blocks below. Don't put any mortar on the cross webs.

After predampening the footing, pick mortar up with your trowel and dump it over the footing. No need to be neat. Spread enough mortar for one corner block. Start where the first corner block is to go. Furrow the mortar by drawing the point of the trowel through it, as in bricklaying. The purpose of this is to concentrate mortar along the face shells of the blocks.

CORNER BLOCKS

Lay the corner block. Line it up above the footing right where your dry layout indicated. Tap the block to bring it down to level and grade. Do this with a level on the block lengthwise, then crosswise. Tap with your mason's hammer or trowel, as professionals do, you'll likely splatter mortar all over your hands. The hammer approach is much neater. Don't hit the block on an exposed corner. You might chip it. Looks bad in the wall.

Gradually work the block down to its ⅜-inch height. Make a story pole or use a measuring rule, whichever is easier for you. If you use 8-inch-high blocks, the top of the first block should be tapped

To cut a block, chip a line across core on both sides with hammer until it breaks.

Keep blocks under wraps so they stay dry. Use a polyethylene sheet or a tarpaulin.

down to 8 inches above the footing. This will leave a ⅜-inch mortar joint below.

Spend the necessary time getting your corner blocks properly placed. The others in the course will be laid according to it. If it's off, they'll all be off. Level first along the length of the block, then across its width.

If the block settles too low while you're working it, pull it out, salvage the mortar and start over. This takes real character. The tendency is to say, "Close enough, the heck with it."

Leftover mortar around blocks after they're laid, can be sliced off and returned to the mortarboard if its clean. Don't try to re-use mortar that spills on the ground or scaffolding. Trowel-mix all salvaged mortar to make it workable.

Lay two corner blocks on the footing in each corner. Square them with a carpenter's square. Lay one more corner block on top of them. Do this at every corner. Each succeeding course of corner blocks is laid so the blocks overlap like fingers on clasped hands producing what is called a *running bond*. You don't have to use running bond, but it makes the strongest wall. Other block patterns may be used if you prefer.

Don't forget that the corner blocks, besides being leveled in two directions, must be directly above the blocks below. Check this by using the level in its plumb (vertical) position. Tap the blocks into place. Plumb first the edge, then the end. Do this with every corner block all the way up the wall. It ensures straight, plumb corners that make a straight, plumb wall.

Professionals lay up four or more corner courses before placing any intermediate blocks. This takes too much skill. We amateurs had best limit ourselves to one or at most two corner courses at a time.

When the corner courses have been completed, lay the in-between blocks. Stretch a line between two corners in the first course. By the time you get the line stretched the corner blocks should be solid enough so they won't move. See that the line is flush with the tops of both corner blocks and drawn tightly enough that sag is imperceptible. If the distance is too great for a no-sag condition, lay one intermediate block and support the line there, too.

Spread beads of mortar for two or three blocks at a time. No need to furrow this mortar. Stand up several in-between blocks on end and butter their end shells. Do your best to keep from getting mortar on the block faces. It saves clean-up later. The mortar will stay on the end faces better if you "wipe" the trowel over the

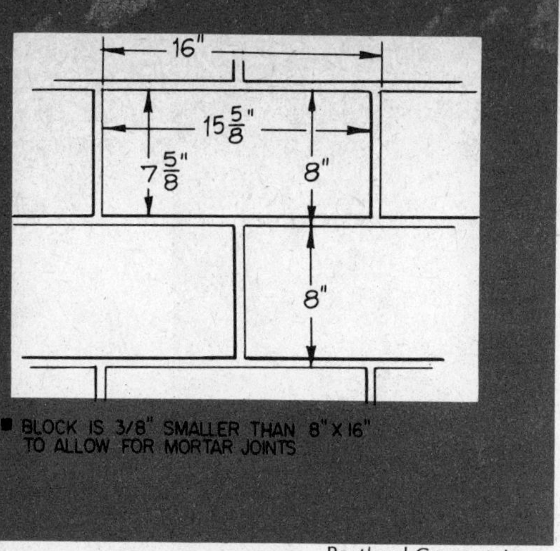

Portland Cement Assn.

A standard 8 x 8 x 16-inch block is actually ⅜-inch smaller in its length and height.

corners of the blocks. Mortar that's too dry, won't stick at all.

Set each mortared block gently on the one below, shoving it against the previously laid block. The side thrust helps ensure well filled joints. As an in-between block is lowered onto its mortar bedding, keep it tipped toward you slightly so you can see the block below it. Spot the block on its bedding with the lower edge directly above the block beneath. By rolling it to a vertical position, and at the same time shoving it sidewise against the previously laid block, it will snuggle up to the line with little fuss.

All adjustments in a block's position must be made while the mortar is still plastic. Later fiddling with the block will break the mortar bond.

See that the end joint (between blocks in the same course) closes to ⅜ inch, or the distance indicated by your dry layout. Then put a level crosswise on the block and tap the block down even with the stringline and leveled. While in-between blocks should be laid flush with the stringline, none should touch it. That throws the line out of whack.

Slice off excess mortar from the joints and go on to laying the next block.

When you get to the last block in a row, a special laying procedure is needed. This block is called a closure

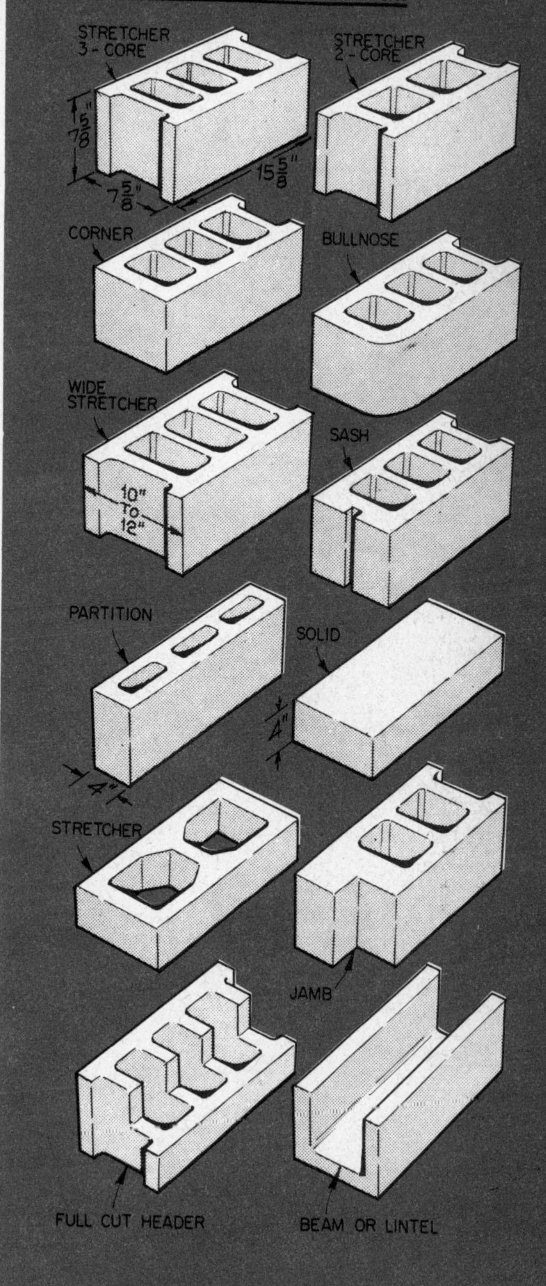

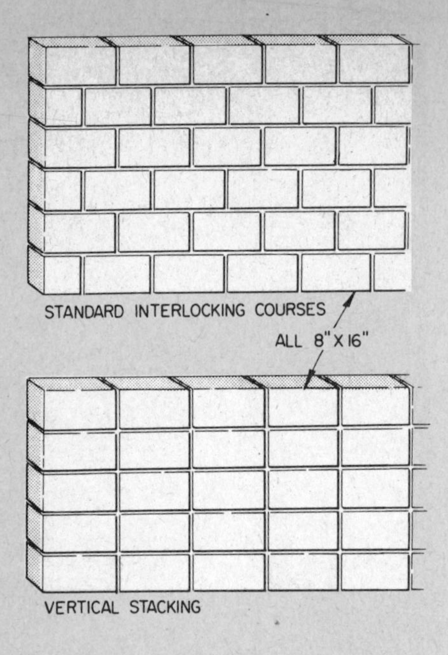

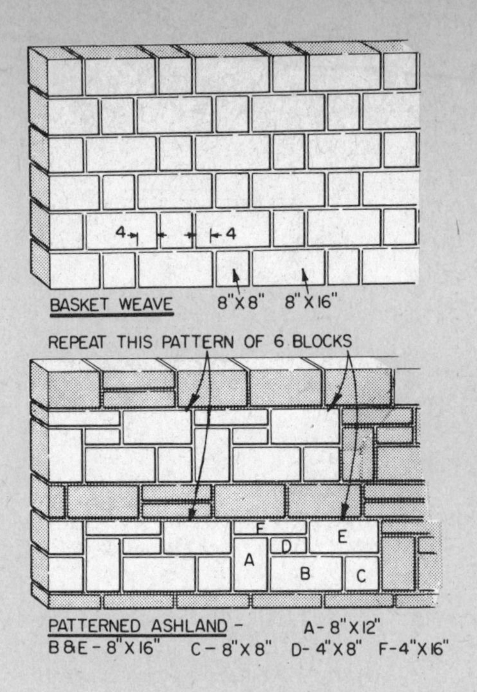

block. Lay it by buttering all end joints with mortar. Put four ribbons of mortar on the block itself and four on the adjoining blocks in the wall. It's a roughie, as you'll find when you try it. It may seem impossible, but you can do it. Don't be too proud to use your fingers if they help.

Lower the buttered closure block into place. Be careful or your painstakingly installed mortar will slough off. If you goof, clean up the mess and start over. Do not stuff mortar into the end joints after the block is in place. That makes a weak spot in the wall.

Don't spread mortar more than several blocks ahead. It tends to stiffen and lose plasticity. In hot, dry weather, spread only one block at a time to keep ahead of moisture loss in the mortar.

CONTROL JOINTS

Concrete masonry shrinks and expands with temperature and humidity changes. If your wall approaches 20 feet in length, it will need what are termed *control joints*. They control cracking in block walls the same as they do in concrete.

A control joint is simply a vertical joint from top to bottom of the wall. Stresses that accumulate in the wall relieve themselves at the control joint before a crack can form. Lateral support must be provided across a control joint. For that reason the use of special control joint blocks is desirable, but not mandatory. You can make a good control joint without them.

Ideally, control joints should create wall sections that are roughly square. In a 4-foot high wall, it would be nice to have a control joint every 4 to 6 feet. Then you could be sure of no cracks. The taller the wall, the farther you can space out control joints.

Likewise, a control joint is needed

Special fence blocks come with pilaster blocks which give rigidity and decoration.

Portland Cement Assn.

Split blocks look like stone masonry when laid into wall. Patio is of blocks, too.

wherever the wall's cross section changes, as from 12 inches to 8 inches thick.

The logical place to put a control joint is at a window, door or other weakened section of the wall. Don't use control joints in basement and foundation walls. They are less exposed to temperature and humidity changes and full strength is needed. Let 'em crack, if they must.

Control joints are laid up in mortar the same as vertical joints in other blocks. Before it hardens, the mortar is raked out to a depth of ¾ inch. This joint is later calked with an elastic compound that will stretch and compress with joint movements. Prime the wall as recommended by the calk manufacturer.

WALLS THAT INTERSECT

Where two walls join, a point of restraint is formed. You may get a crack there unless a control joint is provided. When both walls are bearing walls—that is support weight from above—a ¼×¼-inch steel tiebar is embedded in the joint between the two walls. The tiebar should be 24 inches long and have two right-angle bends at the ends. These are grouted in mortar.

Nonbearing intersecting walls need control joints, too. The design can be simpler, however. Embed a length of ¼-inch hardware cloth in every other course across the intersection. Place the mesh as you build the first wall and leave it projecting until the other is built. Then bend the mesh down into the joint as each course is laid.

Bridge over door and window openings with lintels, such as 3⅝×3⅝-inch steel angles placed back to back. Lay 4-inch thick blocks on them to make an 8-inch wall. Using this method you don't sacrifice the block texture of the wall.

Precast concrete lintels can be made or purchased.

If the wall is to support a frame roof, embed ½×18-inch anchor bolts every 6 to 8 feet along the top two courses. To do this, grout the cores. Place crumpled-up newspapers or wire mesh two courses down in each core to retain the grout. When the wall is completed, 2-inch wood plates are bored to fit over the anchor bolts. Nuts draw them snug.

Window sills often are made of precast concrete or concrete bricks laid on edge. Some building codes call for a continuous concrete bond beam around the top of the wall. Check yours.

THE UNITS

Concrete masonry units fit in with any architectural style—French provincial, contemporary and others. Varied sizes, shapes and textures of blocks can be used to achieve any effect you want. Besides a variety of units to choose from, you can lay blocks in a variety of patterns. Concrete blocks are made in all parts of the country. To get an idea of what's available, visit your concrete products producer. Building materials dealers, who sell blocks but don't make them, may not have a full selection.

Blocks come in standard weight, lightweight and heavyweight. Most used are the lightweight ones. These are made of portland cement and lightweight aggregates such as cinders, pumice, scoria and expanded slag, clay or shale. A lightweight unit usually weighs 25 to 35 pounds.

All blocks are made to comply with ASTM specifications. The specs for hollow load-bearing units require compressive strengths of five blocks to average 1000 psi. of gross area. To meet this requirement one standard block must hold up 64 tons.

When you buy them, concrete blocks should be well cured and dry. Cover them to keep them dry until laid in the wall. Wet blocks, if laid in a wall, will shrink and make cracks.

Blocks are classified as solid or hollow-core units, solid ones have a core area of 25 percent or less of the gross cross-sectional area. Hollow-core blocks are the most common. Units are made in two- and three-core, design. Two-core is more popular in the West. They insulate better because there are fewer webs to transmit heat.

Concrete blocks are made in 4-, 6-, 8-, 10- and 12-inch widths. Heights are 4- and 8-inch. Half units, corner units, jamb units, sash units and many other shapes are to be had.

Block sizes are actually 3/8 inch shorter than their nominal dimensions. For instance, the most commonly used concrete block is the 8×8×16-inch unit. Measured, you'd find it only 7⅝×7⅝ ×15⅝ inches.

The thickness of walls that support roofs or floors is usually 8 inches. Some free-standing walls can be made of 6-inch block. A partition wall carrying nothing more than its own weight can be 4 inches thick. Partitions supporting floors or roofs should be 8 inches thick. Foundations should be made with thick blocks, preferably 12-inch. However 8-inch blocks may be used for a foundation wall less than 4 feet below grade. In light soils this may be extended to 7 feet. In heavy soil with a drainage problem, never build a basement wall of concrete blocks. You won't be able to keep water out. Cast-in-place concrete is preferable. In light, sandy soils, block basements are fine. Use 12-inch blocks depending on the strength needed, and whether frame, solid masonry or masonry veneer is to go above them. If in doubt, always use 12-inch blocks. They're none too heavy for most basement walls.

Concrete bricks are made with the same materials as concrete blocks. They're terrific for piecing between blocks to obtain nonmodular dimensions. Bricks come in two sizes: standard and modular. A standard concrete brick measures 3¾×2¼×8 inches. A modular concrete brick measures 3⅝×2¼×7⅝ inches.

GLAMOR BLOCKS

Besides plain blocks, special ones are built for beauty.

Split block—Probably the queen of them all is the split block. Its texture resembles a rough stone in appearance. Split block's looks in a wall is hard to tell from the most costly stone masonry. Made by breaking a hardened solid concrete unit lengthwise into two pieces, split block is usually pigmented. Colors are available to suit any preference. Split blocks are made in brick thicknesses and are laid similarly to brick. Never wet them before laying. Keep them dry always.

Slump block—Running split block a close second in beauty is the slump block. Unusual treatments, such as adobe effects, are possible. Slump blocks are made with a concrete mixture that is wet enough that, when jolted, it sags or slumps immediately after the factory's mold has been removed. The resulting units vary in height, surface texture and general appearance. They're usually colored in earth tones.

Glazed-face—These units are used because of their easily cleaned sanitary surfaces. They're made by glazing one side of a standard concrete masonry unit. Using them in kitchens, bathrooms, showers and hallways is a practical way to get beautiful, sanitary, yet economical walls.

Ground-face block—A ground-face block is made pretty much like a standard one except that aggregates containing varicolored pieces are used. After curing, the surface is passed over with a grinding wheel to expose the ags just behind the surface. Ground-face units are not widely available yet, but they're catching on. You can get them ground on one face, two faces or two faces and one end. The cost is about double that of a standard unit. The smoother, prettier look may be worth it.

Grille block, screen block—Screen block walls have become popular, especially in year-round warm climates. Phoenix, Arizona—the block capital of the world—has them everywhere. Screen blocks are made in many shapes, sizes and patterns. One manufacturer lists more than 180 varieties. As garden walls around patios, screen walls provide privacy in addition to architectural style.

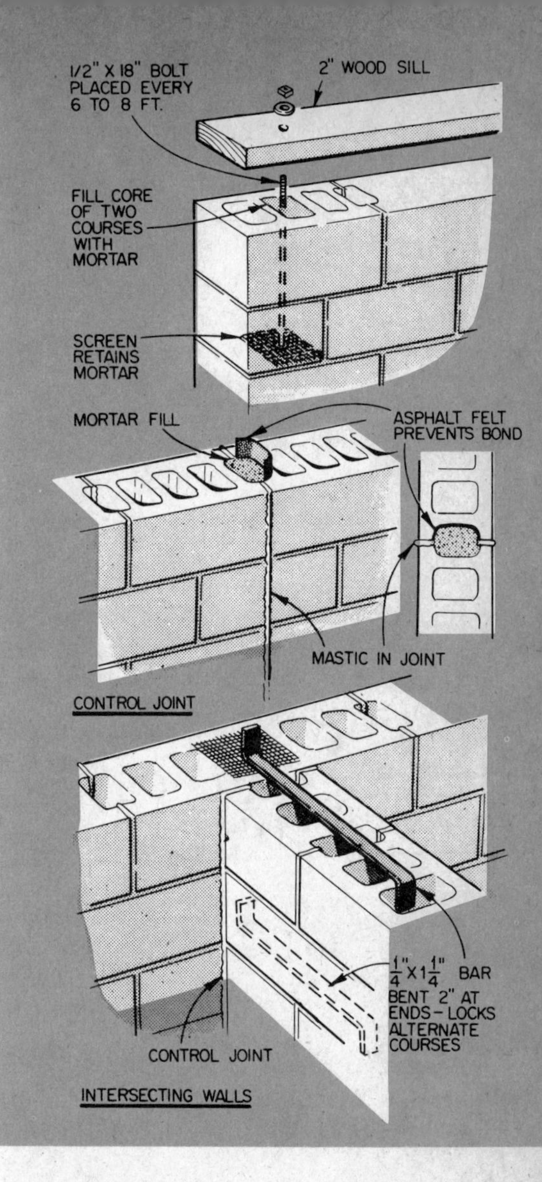

They also let a certain amount of breeze through their openings.

The gardening enthusiast can emphasize specimen plantings with a tasteful backdrop of a grille block masonry wall. Storage bins, fuel tanks, utility meters, even garbage cans, can be concealed behind a screen wall panel. Screen walls can even be used inside the house as room dividers or to set rooms off.

Sculptured block—Some concrete masonry units have an embossed face that creates a three-dimensional effect. Many patterns are available. Check several

If steel mesh is used in joints for added strength, lay it before placing mortar.

If using vertical reinforcing rods, cast them in the footing to align with cores.

Bend rebars down into block core of top course, then grout the core with mortar.

A nonbearing wall is tied to a bearing wall with wire mesh in every other course.

Control joints between bearing walls are held with steel tiebars grouted at ends.

Slip felt into core of standard block and half block and grout for a control joint.

Portland Cement Assn.

Half-high blocks give a brick look to wall. Sculpturing made by hand-shaping mortar.

block producers before making your choice.

Fence-block—A number of concrete products producers make special units designed for building low walls and fences. The supporting columns lay up. Slots in them hold the in-between blocks allowing some expansion and contraction without cracking. One such patented wall has units that can be swiveled to any desired angle to make zigzag, planter squares, hillside retaining planters, etc. These units are held with metal pins driven into the ground. No footing is necessary. No mortar either.

No-mortar block—Another recent development isn't glamorous at all. But it is great for fast wall-building. The units are ground to exact size so they can be laid into a wall without mortar. Reinforcing steel is incorporated into the structure. Then the cores of the blocks are filled with grout to make a solid 8-inch-thick concrete wall.

BEAUTY IN STONE

Wall pattern is provided by the color or shape of the rock used

Stones that are quarried from the earth or found lying on top of it, can be used to build anything that can be made with bricks or blocks. The big difference is that there are no standard sizes or shapes of stones. Every stone does its own thing. Sizes vary from small to large. Shapes vary, too. Stonework goes much slower than brick and block because you have to take time out to shape and fit each piece. To save all this fitting, you can order your stones cut to size. Then you can lay them more like other masonry.

Cost is a factor in the selection of stones. The farther the stones must be brought, the higher their cost. Before you design your first stone masonry project, visit your building stone dealer and look at his samples. Get prices for the quantity of stones needed to build your project. Compare costs with bricks and blocks. You'll usually find that stones cost more. If you're willing to come up with the extra money, go ahead and order.

There is a wide choice of wall bond patterns when building with stone. All are generally classed as *ashlar, squared-stone* or *rubble*. The class depends on the amount of care used in shaping

Stones that are laid around a prefabricated metal barbecue make a rustic natural grille.

Using ready packaged mix is easy. Dump in wheelbarrow, scoop out a depression.

Pour required water into the depression. A workable mortar is your ultimate aim.

pieces of stone and the closeness of the joints. In ashlar masonry the joints are held to ½ inch or less. Depending on the exact arrangement of stones within an ashlar wall, this class of masonry is further classified as *range, broken range* or *random.* Joints ½ inch thick are easy to make, and common. You needn't sweat them.

Rubble masonry can be laid *coursed, uncoursed* or *random coursed,* depending on the appearance you want. This also depends on the size and shape of your stones and the variety of sizes. See what you have to work with before you decide about what coursing to use.

Squared-stone masonry falls between ashlar and rubble in fit of stones. The stone faces are cut square or rectangular so they fit fairly closely when laid with ½- to 1-inch-thick mortar joints.

LAYING STONES

Here are some rules for working with the various classes of stonework:

Stones are laid in mortar the same as other masonry. Like the others, a stone wall should be built on a suitable footing. The pieces should be cleaned and dampened before being laid. Spread a full bed of mortar and lay the stones on it. All trimming and cutting of stones should be done before laying them. In fact, each stone should be fully fitted to its spot in the wall before you spread its mortar bedding.

In ashlar masonry made with weaker types of stone, such as sandstone and limestones, the maximum length of each stone should not be more than three times its height. With stronger stones, such as granite, the maximum can be extended to five times height. Maximum width should be from one-and-one-half times the height in weaker stones to three times the height in stronger ones.

With rubble masonry you needn't worry about exact placement of stones in the wall. However, each joint must be completely filled with mortar. You can prop stones in position with small stone chips to hold while the mortar sets. Chips also may be used to help fill up joints. To build a strong wall, every stone should have a full mortar bearing at all points.

Stratified stones, such as slate, always should be set on their natural beds, not on their edges.

Mix materials thoroughly. Add water if needed until mortar has all it can take.

Spread mortar for a stone, position it, then tap into place with trowel handle.

Place long level across high point of two stones, complete setting by tapping.

A hammer with an axe-like head is used to nick edges for a coarsed rubble wall.

Sakrete photos

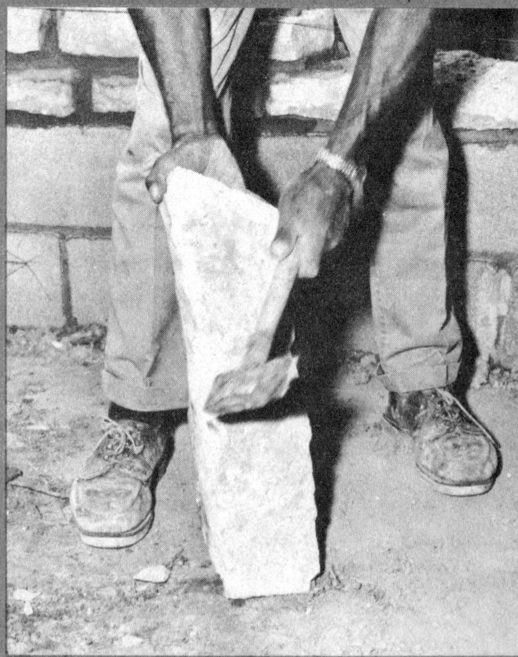

Scratch away 1/4- to 1/2-inch of surface mortar in joints to make stones stand out.

Sakrete photos

Then dry-brush the face of each stone to dust off mortar deposits before setting.

In rough-coursed stonework it isn't practical to make neat looking mortar joints at the face of the wall. Simply rake the joints out about an inch deep and pack mortar in with a narrow pointing trowel. Strike the joints when ready. Use a tool just like for tooling brick and block joints.

The amount of mortar needed in rubble and squared-stone walls varies from about 15 to 35 percent of the volume of the wall. For ashlar walls the amount of mortar depends on the sizes of the stones. The larger the stones, the less mortar is needed. Usually ashlar with ½-inch joints in courses 15 to 20 inches high calls for from 4 to 7 percent mortar. The stones themselves are figured in tons, usually. How much wall a ton will build depends on the type of stone. Better yet, ask your dealer.

TWO-TIERED WALLS

Stone masonry may be used as a veneer over cheaper materials, such as a frame house wall. Stone masonry is often backed up with a tier of concrete blocks. The stone tier is exposed to view; the block tier is hidden in the wall. The two are usually tied together by crimped galvanized metal strips embedded in the mortar joints. Stone veneer laid over a frame wall is tied to the framing similarly, but the ties are nailed to the house wall.

You can use different classes of stone masonry for facing and backing a wall. For example, an ashlar face may be backed up with rubble. Where the backup isn't exposed to view, concrete block is a more practical material.

Just as a brick wall needs headers to tie the two tiers together, a two-tiered stone wall needs either headers or metal ties. If headers are used, it can be done by placing an occasional stone crosswise of the wall, full width. Any stone with its greatest dimension perpendicular to the face of the wall is a header. One with its greatest dimension parallel to the wall face is a stretcher, just as in brick masonry.

The structural bond of a stone wall with headers is achieved by proper arrangement of headers and stretchers. The strongest wall comes from using an equal number of headers and stretchers. Overlap the vertical joints in the two tiers so that no vertical joint is left clear through the wall front to back. The overlap should be at least 4 inches.

Ashlar bond makes the strongest wall. A granite ashlar wall will support about 40 tons per square foot. Hard limestone will support some 35 tons. A square foot of sandstone masonry will hold up about 25 tons. Coursed rubble walls, on the

Stones that are cut and ready to lay in the wall cost more, but you can save time.

A two-tier stone wall is backed up with concrete blocks, bridged by metal ties.

Random rubble wall uses stones that have the faces cut flat for a smoother effect.

A coursed ashlar stone retaining wall is made zig-zag for more lateral strength.

other hand, are only about half this strong using the same materials. Uncoursed, they have still less strength. If you're building a load-bearing retaining wall or foundation, rubble stone masonry is bad news. For a free-standing garden wall or a low planter wall, it's fine. A rubble wall should be at least 4 inches thicker than a similar brick or block wall, never less than 16 inches thick. Ashlar or squared-stone masonry can be the same thickness as would be needed for a brick or block wall.

FAULTS

Stones that aren't carefully selected for wall-building may contain faults that make them undesirable. For instance, granite may contain *knots, sap* and *shakes*. Knots look much like knots in wood and are lighter or darker than the stone around them. Sap is a hard-to-remove surface stain. Shakes are small cracks or seams that may not look bad at first, but often grow larger and more objectionable upon weathering. Sandstone may contain soft seams or soft spots that can be picked out with a nail. Reject such stones.

Limestone with the following faults should be turned down: wide variation in color, uncemented seams, pieces of flint at the surface, cavities, porous areas, discoloration, veins of calcite, and unsound stone. You can test for uncemented seams by soaking the stone a few minutes, then wiping it dry. A wet streak will be left along the bad seam. Cavities, porous areas and unsound spots all affect a stone's structural qualities as well as its appearance.

FLAGSTONE

Stones quarried in flags make good looking sidewalks, patios, entryways and the like. For the best results, flags should be laid in ½ to 1 inch of mortar spread over a 4-inch-thick concrete slab. You can spread the mortar more evenly by striking it off with a 2-foot 1×6 with nails protruding from one edge the same distance as the mortar bed thickness.

Order your flagstones 3/8 to 3/4 inch thick. Buy enough to cover 10 percent more area than necessary to allow for waste in cutting and fitting. Cut the stones roughly to shape with a hammer and mason's set, or chip at the edges with the head of your mason's hammer. Spread mortar for one or two flags at a time. Set them into the mortar by tapping lightly with a rubber hammer. If the flags are smooth enough, a level can be used to help get an even surface.

Fit each flag as you go. Dampen it before setting. Try to use up the small pieces left over after fitting a larger piece.

Rake out all mortar from the joints between flags before it hardens. Later fill the joints slightly higher than the surface. Trim flush after the mortar stiffens somewhat.

Flags are most often made of limestone or slate. They're usually sold by the pound. Use a clear sealer like that recommended in the chapter on getting color into concrete. The sealer will protect your work from staining and make the stones easier to keep clean.

Flags also can be set on a sand bedding or even on a scooped-out hole in your lawn. Don't expect as refined a surface as if you'd laid them on concrete. They are sure to move around with variations in ground moisture and frost.

CLEANING STONEWORK

Take special pains to keep mortar stains off the exposed surfaces of the stones as you lay them. An acid wash can't be used to remove mortar from stones. Acid discolors the stones more than mortar does. Wire-brushing sometimes works. A rotary wire brush chucked in an electric drill is the easiest to use. You'll have to abandon the method if it leaves specks of metal on your stones. The metal flecks will rust and cause later staining.

The most successful all-round method of cleaning stonework is scrubbing with laundry detergent and warm water. Give the wall two washings. Rinse thoroughly after each with a stream from your garden hose. Add household ammonia to the wash water, if you wish, to give it extra "bite."

PREVENTING MASONRY FAULTS

When serious, some concrete defects could condemn a whole project

When a masonry wall cracks, loosens at the mortar joints, develops open joints, soft mortar or efflorescence, something serious is wrong. It's not too late for repairs but it is too late for prevention. The time to have done that was when you built the wall.

The prevention and cure for each fault that masonry commonly develops are described here. Use the one that applies.

Cracks—Masonry walls are continually changing dimensions with variations in temperature and humidity. Cracks, while unsightly, are rarely serious. Sometimes they are caused by uneven settlement of a footing or by one wall being restrained against another. Settlement cracks and restraint cracks are usually wider than shrinkage cracks. They can be serious.

To prevent cracks, build masonry walls on quality concrete footings below the frost line. Make the footings twice as wide as the wall and as deep as the wall is thick. Long walls should be provided with control joints if you don't want cracks to show.

The cure for a mildly cracked wall is patching. Really bad cracks can even call for a new wall to back up the damaged one.

No Bond—Lack of bond between mortar and masonry unit allows the two to separate. Test for good bond as you build the wall. Lay a unit, then pull it out immediately. Mortar should cling to the unit's surfaces. To get good bond, high-suction bricks should be wetted before laying them. Stones should be dampened. Don't wet concrete masonry units, though. Lay full mortar joints in bricks, stones and solid concrete blocks.

There isn't an easy cure for no bond. You must rake out the affected joint and stuff new mortar into it.

Open head joints—Caused by incomplete filling of end joints between masonry units, open head joints make a weaker, less watertight wall that soon needs repointing. Be sure to butter the ends of bricks, blocks and stones well before you lay them. Don't count on slushing the joints full after the unit has been laid. That's poor practice.

Cracks.

Open head joints should be repointed to fill them.

Soft mortar—To withstand weathering, mortar should be hard. If it comes off on your fingertip when rubbed or readily abrades when scraped with a key, the mortar is too soft. Mortar like this most often comes from using too much sand in the mix. Mix your mortar according to the directions in the chapter on

No bond.

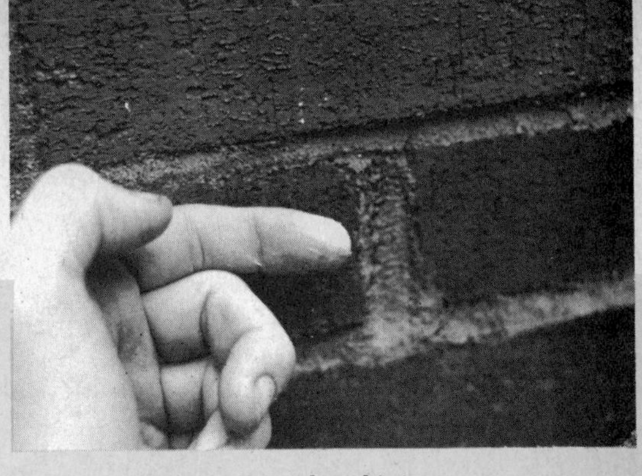

Open head joints.

Soft mortar.

Efflorescence.

mortar-making tips. Discard any mortar more than 2-1/2 hours old. Tool the joints to compact them at the surface.

You can prolong the usable life of soft mortar by applying a good masonry water repellent to the wall. Soft mortar joints will eventually need repointing before the wall is seriously weakened by disintergrating mortar.

Efflorescence—When water gets inside a masonry wall and washes salts in either the mortar or masonry units out onto the surface, the resulting stain is called *efflorescence*. You can prevent it by mixing mortar correctly, protecting the wall from water and using quality units that are free of soluble salts. Tops of exposed walls should be sloped to drain water off, or capped. Build in metal flashings to keep rain water on the outside of masonry house walls. Cover the wall while you're not working on it, so it stays dry.

Efflorescence can usually be removed by a dilute muriatic acid wash followed by thorough rinsing. Dampen the wall with water before you start. Stone masonry and certain kinds of brickwork will have to be cleaned with soap and water rather than acid.

Metal strips were used for forming the colored pebble concrete entry-walk "islands."

MAKE CONCRETE COLORFUL

Use of color sets a project apart from the gray concrete around it

With all the ways there are to put color into concrete, it's a wonder you see so much plain gray concrete around. Convincing someone who has experienced the thrill of making colored concrete, to build the gray stuff would be like trying to switch the owner of a color television set back to black-and-white. Colored concrete is adaptable for any exposed-to-view use.

The methods available for making concrete colorful are these: (1) color integrally by adding a color pigment to the mix; (2) dust color pigment on the surface during finishing; (3) stain the hardened concrete surface; (4) paint it; or (5) use colorful aggregates and expose them. Numbers one and five are the most lasting and effective methods. They're also the most costly in time and effort. Painting is the easiest, cheapest method. Integral and dust-on coloring must be done when the concrete is cast. The other methods apply to finished, cured projects.

INTEGRAL COLORING

Lasting integral color in concrete is obtained with mineral coloring pigments. Synthetic mineral oxides are the best ones, since they have a high degree of coloring power and don't fade. Natural oxides cost less, but are not as pure as synthetics.

You get more for your money with the synthetic colors. All of them come as fine powders that are added to the mixing drum with the cement and other materials. The synthetic oxides are chromium oxide (green), cobalt oxide (blue) and iron oxides, which produce shades of buff, beige, brown, maroon, red and black. The earth colors are cheapest. Green costs more. Blue is so expensive it's limited to small projects. Pigments can be mixed to get in-between colors.

You can't always judge a pigment's coloring ability by its appearance when dry. Use white portland cement with colors if you want clean, bright colors

The bright glare of plain concrete is subdued by adding color to pool deck.

Robert Cleveland photo

or pastels. Use gray portland cement for making darker, more somber colors or for making gray and black. As a rough starting point use 1½ percent coloring pigment in white cement for making pleasing pastels. Use 7 percent coloring pigment for making dark colors. Don't use more than 10 percent coloring pigment. With black, normally half the amount of pigment produces the desired shade.

Colored concrete can do as much as landscaping toward beautifying setting in area around your home.

Right, color possibilities are endless with color mixed in concrete and colorfully exposed aggregates.

Below, section off wood forms to make slabs small enough to handle. Later you can leave the forms in.
Portland Cement Assn.

Always weigh out color pigments. Never batch them by volume. If you do you may get uneven colors, batch to batch. The other materials in colored concrete should be batched carefully, too, in the interests of uniformity, especially the water. Mix colored concrete longer than normal to get complete blending of the color.

Aggregate color—sand and stones—shows through surface color. To avoid sandy undertones use white silica sand in place of your regular sand.

MAKING SAMPLES

A great way to find the correct pigment proportion is to make samples. Weigh out the materials carefully on a postal scale.

First make the colored cement for a sample. Put an ounce of color pigment on the scale. Then, for a 1½-percent pastel add 4 pounds 2 ounces of cement. For fully saturated 7 percent color, add only 14 ounces of cement. Mix the cement and coloring pigment together thoroughly.

With the colored cement you're ready to make a sample batch of colored concrete. If you normally use a 2½-cubic-foot batch as shown in a table in the first chapter of this book, use the amounts opposite the "2½ cu. ft." figure. Divide them by 10 and take them as ounces not pounds. This gives a 1/160th-batch sample. With wet sand, weigh out 5.2 ounces of colored cement, 2.2 ounces of water and 11.8 ounces of sand. Omit the gravel and air-entraining agent, since you're only making a sample. Use them in the final mix, though, if called for. Guess at the tenths of ounces. It's close enough.

Put all the materials into a large can and mix them with a small paddle. Handy forms for casting samples are 1-inch-long cutoffs of 3-inch-diameter plastic drain pipe. Slit each piece up the side so it can be peeled off the sample when hard. Drying of samples can be speeded by placing them in a warm oven for several hours. It ruins them as concrete, but they're fine for color comparisons. In fact they look almost good enough to eat.

Make new samples to test color adjustments using the adjusted amounts of color pigment. When the desired percentage of pigment is established, apply it to the amount of cement you normally use for a batch of concrete. For example,

Always proportion the color pigments by weight. A baby scale is the size needed.

To make color comparisons, cast samples one-inch thick using plastic drain pipe.

Integrally colored concrete with exposed stones was cast in holes dug in the lawn.

if you want 5 percent pigment and use 52 pounds of cement, you'll want 2 pounds 10 ounces of color in each batch of concrete. Weigh this much pigment out into small grocer's bags and staple the tops. Tear one bag open and dump its contents into each batch as you mix it.

Colored ready mix can be ordered for about $3 to $6 per cubic yard more than regular ready mix.

Never use anything but a concrete coloring pigment in concrete. Colors for paints are not suitable. Neither are fabric or food dyes.

Cure with sprayed-on curing compound or clean, wet sand spread over the surface. Dirty sand may stain. Don't try to cure colored concrete with polyethylene sheeting. Wrinkled, it causes spotty colors.

Avoid over-troweling of all colored concrete surfaces. One, two at most, floatings and one light steel-troweling should be the maximum. The less finishing the better.

Good-performing color pigments are made and distributed by Pfizer Minerals, Pigments & Metals Div., Chas. Pfizer & Company and by the Frank D. Davis Co., among others. If you can't get what you want from your building materials dealer, you can write to the Davis firm at 3285 East 26th Street, Los Angeles, Calif., 90023. Coloring agents are not expensive.

Because it is wasteful to color a slab, full thick, a two-course color job may be desirable on slabs. Build a base course of normal concrete. Strike it off ½ to 1 inch below the tops of the forms. Scratch the surface to roughen it, or else cast the second course on top of the first while the first is still wet. Color is used only in the second course. Two-course placement makes your color pigments go much further.

DUST-ON COLOR

For really low-cost colored concrete slabs, use the dust-on method in which coloring pigment is troweled into the slab's surface. Only the top eighth-inch or so is colored. Buy a ready-made dust-on mixture or make your own. To make one, use 2 parts white portland cement, 2 parts fine sand and 1 part mineral pigment of the desired color. Use one of the same pigments as described under integral coloring. You'll need about 50 pounds of dust-on mixture for every 100 square feet to be colored. This takes about 6 pounds of pigment. Dry-mix the ingredients before using them. Make enough to do the whole job or you may get color variations.

Place your slab and strike it off as usual. Concrete can be air-entrained or plain. When all the surface moisture has disappeared, sprinkle the first coat of dust-on mixture evenly over the surface. Let it sift through your fingers onto the surface. Then, after a few minutes, trowel it in. An aluminum or magnesium float is excellent for this. Don't float water to the surface or it will cause a spotty loss of color. Follow troweling immediately with another, lighter dusting and trowel that in. Run all joints and edges before and after dusting. Finish and cure with sprayed-on compound, or clean, wet sand.

Ready-made dust-on mixtures in 100-pound bags are distributed nationwide by Master Builders, Cleveland, Ohio 44118, which makes *Colorcron,* and A. C. Horn Div., Dewey and Almy Chemical Co., 62 Whittemore Ave., Cambridge, Mass., which makes *Colorundum.*

CONCRETE STAINS

Another method of getting attractive color into an existing concrete surface is with concrete stain. There are a number of good ones on the market. When used according to the manufacturer's directions, stains are quite satisfactory.

Stained-on color is not bright, but can be strengthened by coating the colored surface with a wax of the same color. Waxes are sold by stain manufacturers with instructions for use. Occasional rewaxing will be needed, depending on the amount of foot traffic. Wax also helps to brighten faded pigment-colored surfaces.

Don't apply stain to any concrete that's less than 6 weeks old. Even *more* curing and aging is desirable. Before staining, the surface should be clean and free of all foreign substances including curing compound, hardeners, soaps and paints. Some stains call for an acid-etch before staining. Others don't. Read the label.

Stains are usually put on in two coats. The amount of coverage varies with surface porosity. Normally, a gallon of stain will give two applications to about 400 square feet. Coverage of wax is 600 to 900 square feet per gallon for one coat. Start staining and get your application practice where it doesn't count. Not at the front door.

The most satisfactory concrete stains are inorganic ones, made of metallic salts. Salts, such as mild iron or copper sulfates or chlorides, react with lime in concrete to produce brown, yellow and green. Applying separate solutions of different salts can produce other colors. For example, ferric chloride, which makes a brown stain, can be coated over with sodium ferrocyanide to make the surface dark blue.

Metallic-salt stains won't hide imperfections such as patched areas. They should never be applied to old concrete. They're ineffective on it. Concrete under a year old works best. Also, don't apply inorganic stains to slabs of varying ages. The colors you'll get will vary. Don't use an inorganic concrete stain over a pigment-colored surface.

Sift dust-on powder evenly over fresh surface and trowel it in with a lightweight metal float.

Follow manufacturer's directions to apply concrete stain. Different colors give flagstone effect.

Floor paint for concrete is easiest to apply if you use a roller at the end of an extension handle.

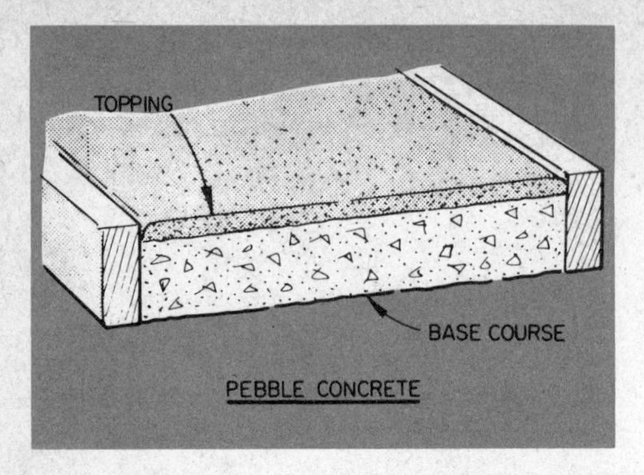

USES FOR COLORED CONCRETE

Garden walls	Basement floors
Wall capping	Tennis and shuffle-board courts
Patios	Parking areas
Driveways	Fireplaces
Sidewalks	Mortar
Stepping stones	Stucco
Swimming pools and decks	Steps

If your dealer doesn't have a good brand of inorganic concrete stain, write to Rohloff & Co., 918 North Western Ave., Los Angeles, Calif., 90029. Ask where you can buy *Kemiko* concrete stain. It comes in eight colors from browns, beiges, greens and rusts to black. *Col-r-Tone,* also available, enhances the color effect both with *Kemiko* stain and pigment-colored concrete. It covers imperfections too, according to the manufacturer. A 9-year life is claimed for a properly stained surface in a mild climate.

Oil stains are commonly used for getting wood-like colors on concrete. These hide imperfections but don't wear nearly as well on slabs as inorganic stains do. Cabot and Rez both make stains for concrete. They may be applied by brush or roller. They're usually available at paint dealers.

You can even use ordinary wood-type oil stains on concrete. If you do, first apply a dilute solution of zinc sulfate (2 pounds per gallon of water). This keeps the alkalis in the concrete from making soap out of oils in the stain.

Aniline dyes can be used to stain concrete in many colors. In wax solutions, they're often diluted with naphtha. These are best for restoring integral or dust-on colors that have become weathered or faded.

All stains disappear after a while. Waxing will improve things. Don't use a wax without a stain beneath it. The results will disappoint you. It's best to buy stains that are sold specifically for use on concrete. There are a number of different brands. Follow the directions for the one you are using.

PAINTS

Your choice of colors is by far the greatest with paints. Oil-base, latex, alkyd, chlorinated rubber, portland cement paints or catalytic coatings can be used on concrete. Latex beats all for most uses. Not for garages, however, where car tires make them lift off. For floors, chlorinated rubber floor and deck paint has the advantage of long wear. For the heaviest wear on floors use one of the catalytic epoxy coatings. Outdoors, they chalk heavily, however. A drawback. Portland cement paints are good for walls, never for floors.

To make exposed-ag concrete, strike off base course of concrete 3/4″ below forms.

Before painting, a surface must be clean and dust-free. Sometimes acid-etching is required. Consult the label on the paint you'll be using. Apply paint to floors with a long-handled roller. It saves your back. For more information on painting concrete, see GUIDE TO PAINTING YOUR OWN HOME (Fawcett Book 646).

COLORED AGGREGATE

Colored concrete made by brushing away the surface mortar to expose colorful aggregate just beneath is called *pebble concrete*. It can be either *exposed-aggregate* concrete or *rustic terrazzo* depending on how it's made. Once completed, the two are tough to tell apart. They give a super, luxury look to your house and grounds.

You can use any aggregate from pebbles to 2-inch rocks and larger, as long as they're clean, sound and otherwise suitable for making concrete. All the stones need not be the same size, but they may be. It depends on what effect you want. Get various samples, if you wish, and try them in 1x1-foot sample squares.

Mixing your own is the best way to make pebble concrete. If you order ready mix, you're stuck with whatever aggregate comes with it. Sometimes it's pretty bad, from the looks standpoint.

Design your pebble concrete project so you can work in small sections, doing one section at a time. Big jobs done all at once require too much manpower.

Don't tackle more than about 50 square feet of pebble concrete at once. Get this much ready for ag-exposure before you start the next. You may leave the form boards in, or butt the sections against each other by removing the forms once the concrete has set. Either way you'll be providing ready-made contraction joints between sections to prevent cracking.

EXPOSED-AG

The best exposed-aggregate concrete is made in two courses—a base course of regular concrete and a topping of plain or colored sand mix concrete. The special aggregate is placed into the top course by "seeding," which concentrates the stones just below the surface. It also conserves expensive aggregates.

Place and strike off the base course for one section about ¾ inch below the

Dump surface course onto still-wet base. The topping may be colored, if desired.

Strike top flush with forms. Note the "roll" formed by air-entrained concrete.

Seed select stones evenly over the soft topping and hand-arrange them, if needed.

Tamp stones into topping just below the surface using a 2x4 or other flat tool.

top of the forms. Before the base course has had a chance to set, mix and place the sand-mix course. Mix about one-fourth as much as was needed for the base course to make a nominal 4-inch-thick slab. Use no stones. Strike the sand-mix top course off even with the forms.

Dampen the surface aggregate and seed it onto the section by hand, shovel or pail. Arrange the stones individually, if necessary, to get them spread out in a single, even layer across the entire section. Give the edges special care.

Tamp the stones into the surface course with a darby, a wood float or the edge of a 2x4. The tool to use depends on how stiff the top course is. If you wait too long before tamping, a brick makes a good emergency tamper. Don't set the stones much below the surface. Ideally there should be just a thin layer of mortar covering each stone. Have some extra stones handy for fill-in where needed.

When you can broom and flush the surface without dislodging any stones, the concrete is ready for exposure. This usually takes about two hours, depending on the mix and the weather. Try exposing a spot gently at first with a stiff-bristled garage broom and slight water flushing from a garden hose. If it goes well, do the whole section. The water wash should take away any film of mortar on the exposed stones. If you wait too long and exposure comes hard, have a wire brush on hand to scour the mortar away.

Don't expose more than the top third of each stone. Better yet, only the top quarter. The more you expose, the more texture the surface will have. Larger ags produce rougher textures. Too rough a texture makes hard walking, especially in high heels. Cure as usual. After curing, an acid-wash will brighten the aggregate.

RUSTIC TERRAZZO

Rustic terrazzo is the terrazzo contractor's answer to exposed aggregate. It's also called *unground terrazzo,* because the aggregates are exposed by brooming, not by grinding and polishing, as with ordinary terrazzo. Rustic terrazzo is nearly always made in two courses with color in the topping. The color may match or contrast with the color of the aggregates.

Rather than seed in special ags into the topping after it's placed, the surface ags in rustic terrazzo are mixed right in with the topping. The aggregates may be marble screenings, purchased from a terrazzo supplier. They can also be quartz, crushed tile, crushed brick, river-washed gravel or any other material that's compatible with concrete and will take weathering. While you could use scrunched-up 7-Up bottles for a topping

Flush and brush mortar around the stones with a hose and a stiff birstled brush.

Use two kinds of stones for two-toned effect. Concrete pigment makes third color.

aggregate, they wouldn't make the best surface for barefoot dancing.

Build rustic terrazzo with a ⅝-inch-thick topping. The simplest method is to cast all of the base course at once. Build the forms to section your project off into 6x6-foot or smaller pieces. This lets you use ready mix for the base. Scratch the base with a notched strike board or by brooming. Let it harden. Topping may be placed before or after the base course concrete has dried out. Any placed after the base has whitened, requires special bonding. For this, scrub a thick cement-water grout onto the base just before placing the topping.

TOPPING MIXES

Topping concrete can vary from ordinary colored concrete made with your selected aggregate to a special terrazzo mix. For a good terrazzo mix see the mix table in the final chapter. Most terrazzo mixes contain cement, water and marble chips, but no sand.

Here is a successful mix for rustic terrazzo using small marble chips. I used it on a patio project to get a pale green mortar color with white and green marble chips. You could use it for any color of rustic terrazzo: White cement—45 pounds; Color pigment—11 ounces; Marble chips—green, 28 pounds; white, 28 pounds.

The color pigment used in this mix was at the rate of 1½ percent. To get fully saturated 7 percent color, you'd need 4 pounds of pigment instead of just 11 ounces. This batch covered an area of about 20 square feet. The proportions can be varied to suit your batch-size.

Strike off the topping and darby or bull-float it, but don't finish it beyond that. Later it can be broomed and flushed when ready. Cure and acid-etch, the same as for exposed-ag concrete.

If any stones ever come out of your pebble concrete leaving unsightly holes in its surface beauty, you can replace them quickly with epoxy resin. Properly built, there should be few dropouts.

When your pebble concrete has cured and dried out, you can recapture the luster of wet aggregate by sealing it with a glossy or nonglossy sealer. Glossy sealers call for heavy coatings and make the surface look wet. Applying them in two medium coats is better than putting on one heavy coat. Pinholes clear through the coating are avoided.

Good glossy sealers are *Thoroglaze H*, by Standard Drywall Products, Inc.; and Preco *EA Sealer*. Good nonglossy coatings are regular *Thoroglaze; Horntraz,* by Dewey and Almy Chemical Co.; and *Terra-Seal,* by Hilliard Chemical Co. These don't yellow, as some of the others do. They help keep your colored concrete colorful and clean.

CASTING FOOTINGS,

Footings, walls and floors are well within the scope of us home handymen provided we keep them small and simple. They should be small to keep from seeming like never-ending projects. Large walls also bring on design and construction problems that should be avoided. Simplicity makes for easy construction. This doesn't mean that your wall designs have to be stark and uninteresting. Just avoid designs that are difficult to form.

Concrete footings are needed for foundation walls, piers, columns, basement walls and for concrete walls and fences of all kinds. The purpose of a footing is to provide solid support for whatever rests on it. Footings distribute concentrated loads over a broad enough area of soil so the soil can bear up under them.

Always cast a footing on slightly dampened, but not wet, soil. Never place footing concrete on mud, frozen ground or back-filled soil. Instead, the bottom of

the footing should rest on unexcavated earth.

Nearly always, you want the top of a footing to reach a certain line and grade—and be level. Cast it of quality concrete mix-it-yourself or ready mix. Strike it off and float once. You don't even need to use forms for footings if you can cast them in the bottom of a trench and get a level enough surface.

Footings are as simple to design as they are to cast. Generally, a footing for a wall is twice as wide as the wall and of the same depth as the wall is thick. In poor load-bearing soils (soft clays, black dirt, etc.) footings that carry heavy loads may need to be spread over a larger area. In that case, steel reinforcement should be added across the footing and about 2 inches up from the bottom. The re-bars can be wired at the right height to short steel rods driven into the ground. Consult your building inspector for local requirements. Some footings, such as those for retaining walls, also should contain re-bars along their length.

Any time you use reinforcement in concrete to make a stronger structure, do it according to an engineered plan. Otherwise you'll probably be way under or way over on steel. Re-bars should be fully tied with No. 16 wire at all joints and splices. Spliced bars should overlap 40 bar diameters. Thus, ⅜-inch bars should overlap 15 inches. Sometimes footings for porch slabs, carport slabs and garages are cast in one piece with the slab. This is called an integral footing-slab. No reinforcement is required. Build it on 4 inches of compacted crushed stone fill full-width.

CONCRETE WALLS

Nothing can approach the concrete wall for strength. Most of the work of building one comes in forming it. Even with plywood form facings and wire ties, form construction can take four times as long as casting and stripping. If a cast-in-place concrete wall is what you want, don't let forming scare you off. It's not beyond the enterprising do-it-yourself builder to form and place concrete for an entire basement of a house. (See *How To Build and Contract Your Own House*, Fawcett Book #627.) The plywood forming materials can then be reused as walls and roof sheathing.

Cast-in-place walls can be massive or light, depending on how you form them.
Robert Cleveland photo

Portland Cement Assn.

Every wall starts with a sound footing. Use quality concrete on a solid subgrade.

In forming for concrete, it is important to brace the form to take extreme pressures developed by the fresh concrete being dumped in them. The easiest way to make forms is with ¾-inch plywood braced with 2x4 studs on 12-inch centers. The braces for a low 4-foot wall are held at the bottom by a 2x4 toeboard and at the top by angle braces nailed to stakes. No. 10 wire ties between opposite forms hold them inward against outward pressure of the concrete. Use intermittent 1x2 spacers cut to the wall width and wedged between the forms to keep them apart. As you place concrete, remove these from the form.

Plywood grades suitable are: ¾-or ⅝-inch exterior-type, A-C group 1 douglas fir. Group 1 is stiffer than groups 2 to 4. Form stiffness gives the best surface. For the lowest cost, ¾- or ⅝-inch exterior-type C-C group 1 douglas fir, plugged and touch-sanded is the plywood to use. With the A-C plywood, use the best side in. The plywood's face grain should run across the studs. Avoid using southern pine plywood, if you can, because of a problem with wrinkling in form work.

Do not try to place more than a 4-foot depth of concrete in one hour without making the forms even stiffer than these specifications. If you don't mind slight wiggles in the finished concrete wall between the form studs, you can space the studs out as far as 16 inches on centers, using ¾-inch plywood.

Curving walls can be cast using plywood forms. Saber-saw a set of ¾-inch plywood ribs to the proper curvature. Place the ribs every 12 inches up the curve, eliminating all 2x4 studs on curved sections.

Don't try to bend a plywood panel

Plywood braced with 2x4's is great for forming walls. Use ¾" or ⅝" plywood.

The top of a wall should be floated and troweled just as though it were a slab.

tighter than the minimum radius prescribed for its thickness (see table). If tighter bends are required, use double or tripled sheets of thinner plywood.

To keep the bent plywood from flattening the curvature of your braces, use this trick: On inside curves clamp the plywood bent against 2x2-inch blocks between the upper and lower ribs and the panel in the center of the curve. Nail the plywood to ribs at the ends first. Remove the blocks and clamps. Nail the plywood to the braces working from the outside toward the center of the panel. This will create a stress in the form to counteract any tendency to deform.

On outside curves nail the ends first with 2x2-inch blocks between the panel center and the ribs. Remove the blocks and nail down the panel working from the ends in.

Brace the form against several vertical 2x4's. Stake them and angle-brace just as you would a flat form. Box-outs for any openings through a cast-in-place wall should be placed in the forms and securely nailed.

If the wall is to resist lateral pressure, "key" the footing by embedding a 2x4 in its upper surface when you make it. As soon as the concrete stiffens, remove the 2x4. Don't leave it until the concrete hardens or you'll have to shred the 2x4 to get it out. Don't place any concrete until the forms are fully braced. Nothing can get you more upset than a collapsed form half full of ready mix and a half-empty truck awaiting your orders at $10-an-hour waiting time.

The forms should be tight against the footings to prevent leakage of concrete there. Oil the form surfaces that will be in contact with concrete. Use special form oil or old crankcase oil.

Textured plywood forms will impart their surface texture to your wall. Rough-sawn, sandblasted, combed and even Texture 1-11 surfaces look good on concrete. To get other textures you can fasten things inside the forms before placing concrete. Textured rubber matting works well for this purpose. Moldings nailed inside the form will give patterns in the finished concrete wall. You can make these as fancy as you wish.

To prevent air pockets from spoiling the surface next to the form, tap form gently.

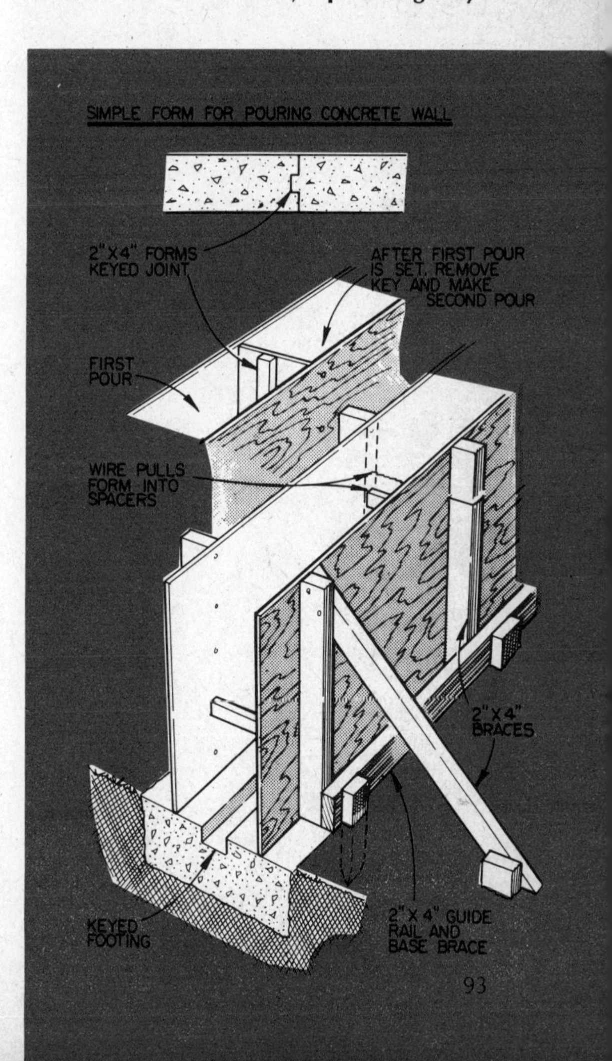

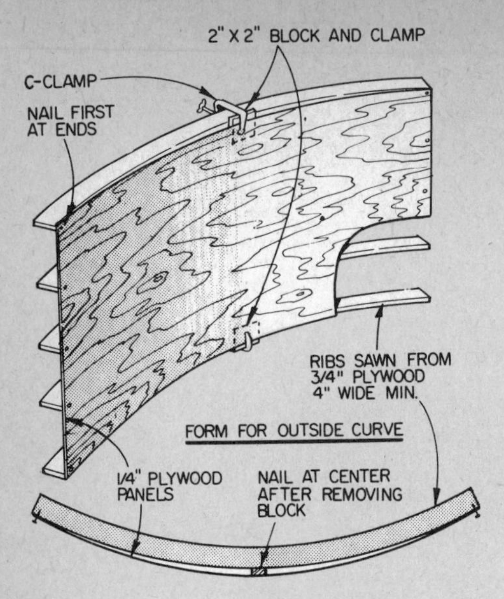

As you place concrete in the form, don't drop it more than a few feet. Place it in 12- to 18-inch lifts. Consolidate by shoveling next to the forms and tapping the sides with a hammer. Leave the forms on until the curing period is over.

RETAINING WALLS

A special kind of wall is the retaining wall. A retaining wall must not only look good, it must hold back tons of force from one side. For that reason every retaining wall should be built with an engineered design. Often, local building codes set standards for retaining walls. Since you must follow these, not too much can be shown about design here.

Sometimes retaining walls are built with what is called batter, that is, the wall face is not vertical but slopes inward slightly. Batter prevents the wall from looking as though it slopes outward. It can be as extreme as 6 inches inward slope in a three-foot wall.

Solid retaining walls should have drains through them for water that collects behind the wall. Usually a 4-inch tile every 10 feet is sufficient.

Retaining wall footings often are offset to prevent wall tip-over. They also may be keyed on the bottom and top to prevent push-away of wall or footing. Sometimes an object called a *deadman* is tied to the wall and embedded in earth fill behind the wall. This helps hold the wall against push-over.

American Plywood Assn.

Concrete molds itself to sandblasted plywood surface, gives a rough wood texture.

Plywood wall forms must be well braced to withstand the pressures from inside.

Portland Cement Assn.

Concrete blocks placed in forms sideways can be used as through-the-wall drains.

Bent ½-inch bolts are inserted into concrete to tie framing above to foundation.

Place bolts at corners and every 6 feet to secure the plate and building above.

One way around a building code that calls for walls over four feet high to be engineered is to build a series of walls less than the maximum height and arrange them up a slope. Such walls are often planted with vines or other cover for a decorative effect.

CONCRETE FLOORS

Two concrete floors you may have occasion to build are garage (or carport) floors and basement floors. Most of the forming, placing, finishing and curing instructions given in previous chapters apply to these floors the same as to slabs. The floor is always the last-built of the footing-wall-floor combination.

A basement floor needs something to keep it from bonding to the footing. The simplest method is to spread one inch of sand over the top of the footing as a bond-breaker.

A garage or basement floor should be cast 4 inches thick. Since floors must slope for drainage, the subgrade under them should be sloped, too, for uniform floor thickness.

An easy way to form the sloping concrete surface is to set out screeds that have the same slope as that wanted in the floor. In a basement floor the screeds should slope toward floor drains. In a garage and carport they should slope toward the doorway. In both, the line around the edge should be level. Establish the level line first. Then place the

On garage footing-slab use tapered strips to form top of footing and sloping floor.

Let floor slope with bottom of tapered strip. Footing is flush with top of strip.

With tapered strip use a wet screed for striking off floor. Create it with float.

Then guide 2x4 strike board on the wet screeds to strike off center of the slab.

screeds to slope away from it (see drawing). Screeds are nailed to 2x2-inch wood stakes driven into the subgrade. To use the screeds, concrete is dumped all around them. Strike-off with a straightedge held at the level line at one end and resting on the screed at the other. After strikeoff, pull the screed out and chunk shovels of concrete in its place.

You can use a unique tapered-strip system to cast a sloping garage floor that has level footings around its edges. The footings serve for attaching wall plate and framing members. Sloped 3 inches back to front, the floor drains well. All framing, except that along the rear wall, is above the level of the floor so it stays dry even when you wash your car. As shown in the drawing and photographs, the tapered strip is held inside the forms by 2x2 wood blocks.

The footing concrete can be struck off across the top of the form and tapered strip. However, there's no strikeoff surface for the sloping garage floor. Therefore, you must create what is called a *wet screed*. This is a smoothed-off portion of floor on both sides of the garage next to the tapered strips. You make it by hand with a float. It should be flush with the bottom of the strip and about two feet wide. A 2x4 strike board, slightly shorter than the distance between the tapered strips, is see-sawed across the wet screeds on each side to strike off the garage floor in between. Wet screeds can be used anywhere that there aren't form tops to use as guides for the strike board.

CONTROL JOINTS

Very few garage and basement floors are built with control joints. That's why there are very few such floors without cracks. You have to live with the floor, so do it right. Space control joints at 10- to 15-foot intervals. Cut them one-fifth the depth of the floor. A two-car garage floor should be jointed into four separate slabs. The joints in a basement floor should come into weakened sections of the floor, such as columns, drains and chimney slabs.

Whatever you do, don't fall for the pitch that steel mesh in the floor will keep it from cracking. That's the old chicken-wire-in-the-floor trick. It'll fool you every time.

If your soil has a water problem, you'll want to build watertight basement walls and floors. Don't count on foundation coatings to do it. They don't work. Good concrete is watertight by itself. It doesn't need them. If the wall cracks, your foundation coating cracks right along with it. Exterior footing tile drains aren't practical either. They soon plug up with soil and no longer work. For the same money you'd spend on coatings and outside drains, you can do the job right. Here's how.

The secret is *inside* tile drains and a material called *bentonite*. Benetonite is a powder, a type of clay that expands up to 15 times its original volume when wetted. If you had a handful of it, you could pour water in a dished-out portion and your hand wouldn't get wet. The expanding bentonite seals off all water penetration. It does the same thing when spread in a thin sheet against your foundation wall. Panels of the material, which has been sifted into the corrugations in cardboard, are nailed to the outside of the foundation. A cove of bulk bentonite is poured along the joint between wall and footing. The benetonite goes to work after the backfill has been placed, preventing seepage through the wall. The few contractors who build basements this way can guarantee them not to leak.

Any water that comes up under the floor is drained away by the underfloor tiles to a sump. Gravity flow or a sump pump dispose of the water.

Don't count on bentonite to prevent seepage in a concrete block basement. If there's a water problem, don't build with block.

If you cannot find bentonite locally, write to the American Colloid Co., 5100 Suffield Court, Skokie, Ill. 60077. Also, ask for directions on installation of bentonite panels.

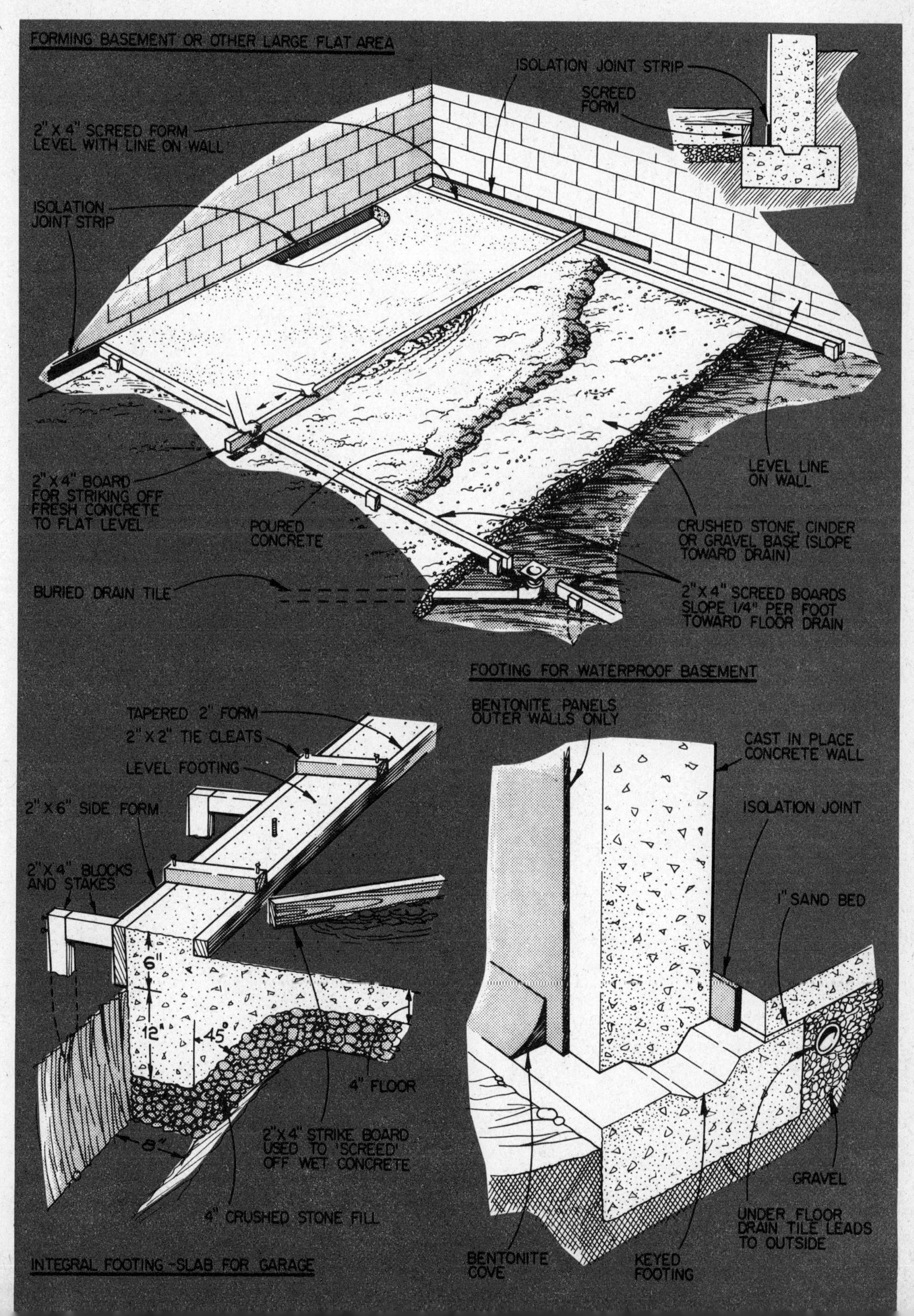

FASTENERS TO USE

There is a fastener for any weight, or any type of wall

Many types of fasteners can be used on concrete and masonry basement walls and floors. The type selected depends on what it's for.

Basically there are five ways of fastening to a concrete or masonry wall or concrete floor: anchoring into a bored hole, driving in nail-like pins or studs, toggle-type bolts, cast in anchors and adhesives.

The drive pins are simplest. They vary from old-fashioned cut nails to modern hardened explosive-driven pins. They hold just the way nails do by friction against the sides of the opening.

Anchors that slip into bored holes and expand hold in much the same way, but first you have to bore the hole. Other anchors are cast into concrete when it's made. They, in effect, provide threaded concrete. Toggle bolts are simple mechanical holding devices that require hollow-core construction. Adhesives are used to stick nails to the wall so that things can be held by them.

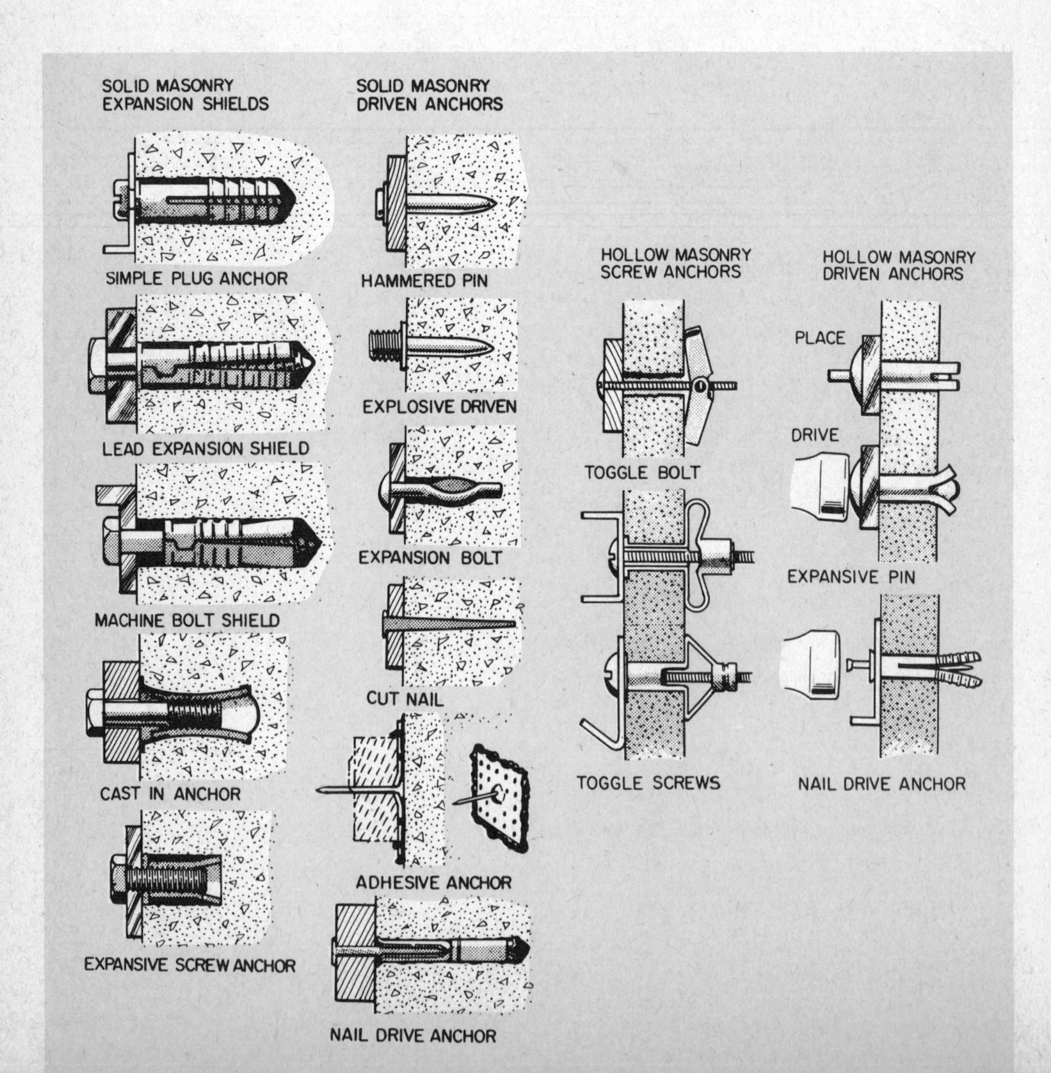

The weight of the object you want to hold and the type of load (shear, pull-out; dead, or live) determine what type of fastener you should use. So does the type of material behind the fastener and the speed of installation. Fasteners that require bored holes naturally take longer to install than those that don't.

FASTENERS FOR WALLS AND FLOORS

FASTENER	USES	INSTALLATION AND DETAILS
Simple plug anchor	Mounting anything to a wall or floor that is normally held with screws.	Bore hole, slip plug in, insert screw and tighten on fixture. Available in plastic, nylon (two types), fiber or neoprene. Plastic and lead plugs do not come with screws.
Lead expansion siheld	Mounting jobs on floors and walls that are too heavy for screws.	Bore hole, push in shield, insert lag screw and tighten on fixture. Use short shields for concrete, long ones for masonry. Not for block unless placed in solid portion.
Machine bolt shield	Mounting heavy items.	Same as for lead expansion shield, which it resembles.
Nail drive anchor	Light anchoring to concrete and masonry. Furring. etc.	Bore small hole, slip shield in, place nail through mounting hole and drive into anchor. Spiral-nail type holds best.
Hand-driven pin (also stud)	Fastening wood or metal to masonry or concrete. Excellent for 1″ and 2″ wood.	Choose pin to penetrate ½″ to 1″ in concrete, ¾″ to 1¼″ in block or mortar. Install with driving tool and heavy hammer. Use a shorter pin if it bends. Fast, cheap, no drilling.
Explosive-driven pin (also stud)	Same as above but faster and less work.	Same as above but pin is driven with 22 cal. explosive charge in a special tool. Get directions and use tool with care.
Expansive screw anchor	Heavy holding jobs in concrete or masonry.	Bore hole, drop in anchor and drive home with simple tool. Insert machine screw and tiahten on fixture. Really holds.
Expansive pin	Holding thin objects to walls and floors.	Bore hole, insert pin through hole in fixture and hammer stud in. Pin expands holding to sides of hole. Light-duty only.
Masonry nail	Mounting wood or metal to walls and floors.	Just drive it in like a nail, which it is. Least expensive fastener for masonry, but often bends when used on hard concrete.
Cut nail	Mounting furring strips.	Drive through furring into mortar joints between blocks.
Adhesive anchor	Applying furring strips, mainly.	Put mastic on wall and squish perforated metal plate into it. Let set. Drive furring over protruding nail and clinch nail.
Expansion bolt	Heavy duty holding in concrete and block.	Bore hole in fixture and masonry. Hammer through hole in fixture into bored hole in masonry. Excellent, low cost.
Toggle bolt	Mounting wood and metal to hollow-core concrete block.	Bore hole into core of block. Slip through hole in fixture and hole in block. Tighten screw to hold. Made in gravity, tumble and spring types. Lost if removed.
Toggle screw anchor	Same as above.	Same as above. Must suit thickness of block wall. Re-usable.
Cast-in anchor	Light or heavy anchoring to cast concrete using machine screws.	Anchor is fastened to form and cast into wall or slab when built.

BUILDING A PATIO IS SIMPLE

An outdoor extension of your home, a place of privacy and comfort

Whether you use a patio as a place to relax, for recreation or for entertainment, it can be a valuable addition to your home. Don't restrict your thinking to a plain troweled-smooth gray concrete patio. This is the kind most contractors will put in for about 50 cents a square foot. Consider, instead, pebble concrete, bricks, concrete tiles or flagstone. If you're going to do the job, you may as well do it with grace.

Many patios are made one-fifth the area of the house. A 1500-square-foot house would get a patio of about 300 square feet, or 15x20 feet square. However, a patio as small as 12x12 feet can be arranged for pleasant outdoor living. A large patio can be extended to include a porch or pool deck.

When you decide where to put the patio, consider access to the house, view, sun, breeze, privacy and drainage.

CONCRETE PATIO

Contrary to what many think, a cast-in-place concrete patio need not be built on a crushed stone or sand subbase. In

Stake 2x4 forms into place using wooden stakes at a maximum of 4-foot intervals.

Prop up planks for a wheelbarrow run to keep from disturbing the staked forms.

Portland Cement Assn.

Plan the patio as an outdoor living room. Access to house, sun, breeze are important.

most cases you can build it right on the ground. The only places you need granular materials under a concrete patio are soft, mucky spots or those with poorly drained soils.

A patio should be sloped away from the house ⅛ inch per foot in order to drain. Avoid getting low spots or "birdbaths" that catch water.

To build a concrete patio, dig out all sod and black dirt in the area of the patio and several inches beyond. The extra digging will make room for your forms. Remove all plants, bricks, wood, large rocks and other debris. Level the subgrade and compact it with a tamper. You can make the tamper of a 5- or 6-foot 2x4 or 4x4 by nailing on a 6x6-inch piece of plywood.

Lay out 2x4 forms and nail them securely to 1x2 or 2x2 stakes driven into the ground at 4-foot maximum intervals. The use of a level and stringline will help you get the forms at the right level and slope.

If you use redwood, cedar or creosote-soaked fir or pine form lumber, the forms can be left in place. They help section it off and add interest to the completed patio. Grids of 3 to 5 feet in size are convenient. The maximum should be 10x10 feet. Leave openings in the patio where you want them for sandbox, planter, pool, etc. Simply form around the opening, placing it next to a control joint.

There's no need to limit yourself to straight patio edges. Long-radius curves can be formed with 1x4 lumber instead of 2x4's. Short-radius curves can be formed with ¼x4-inch plywood or hardboard strips. Cut plywood so the grain runs vertically when in place. This makes it easier to bend.

The old trick of laying out a garden

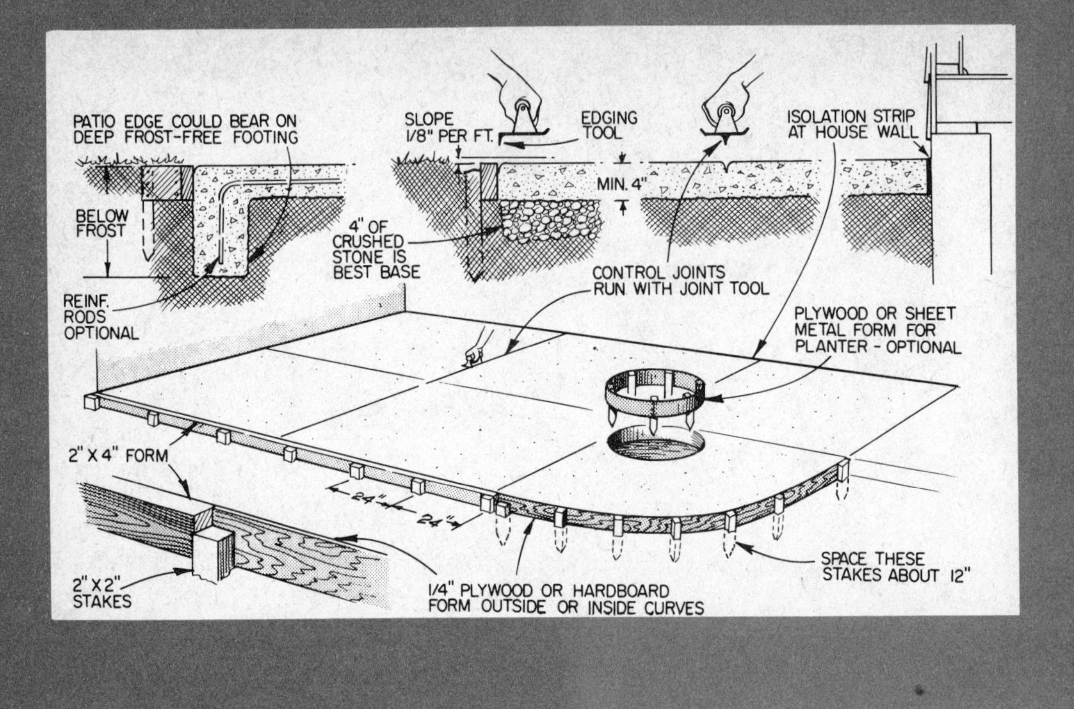

ESTIMATING CONCRETE FOR A PATIO

Area in Square Feet	10	25	50	100	200	300
Cubic yards needed	.11	.28	.55	1.1	2.2	3.3

When finishing patio, remember more than one steel-troweling makes it too slick.

Portland Cement Assn.

hose to outline smooth curves is still a good one. Bend the form around firmly driven *inside* stakes placed according to the hose. Then drive *outside* stakes and nail the form to them. The form will hold its shape after the inside stakes are pulled.

Along curves the form stakes should be placed every 2 feet. Put a 1x4 form stake at every joint in the forms. Position a 2x2 stake at the joint between a curved and straight section to hold the forms in alignment at that point (see drawing).

Place isolation joints at every existing slab and wall. If the special ½x4-inch asphalt-impregnated joint material isn't available, you can use a pair of ½x4 pieces of beveled siding. Remove the siding after the concrete has set and fill the joint with liquid asphalt to keep water out.

Once the forms have been set, use a template to help smooth the subgrade to the uniform 3⅝-inch depth. Soak clay

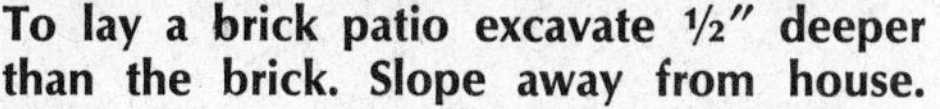

To lay a brick patio excavate ½″ deeper than the brick. Slope away from house.

Frame all around with bricks set on edge in 2″ trench. Align bricks with string.

soil a few days before placing concrete on it. There should be no standing water.

To find the amount of concrete needed for a 4-inch-thick patio cast using 2x4 forms, see the table. For a patio 12x25 feet, for example, first multiply the dimensions and find the square feet to be paved: 12x25=300 square feet. According to the table you'd need 3.3 cubic yards. Order 3½ cubic yards if you use ready mix, or order enough materials to make that much mix-it-yourself concrete.

Air-entrained mix should be used. Place, finish and cure as described previously. Outside edges of the patio should be run with an edger. So should the intermediate edges along forms that are being left in.

If the forms don't do it, the patio should be jointed into 10x10-foot or smaller sections using a jointing tool. No wire mesh or reinforcing steel is needed in a simple patio slab.

If the forms aren't to be left in, they can be stripped, scraped clean and the lumber reused. Avoid trying to saw used form lumber with anything but a carbide-tipped blade. The small amount of concrete left on the wood's surface will tear heck out of your saw blade.

BRICK PATIO

Bricks, make good looking patios, too. The bricks should be hard-burned ones. Use the easy mortarless method. Brick pavers vary in thickness from 1½ to about 2 inches, depending on the manufacturer. Face sizes are about 4x8 inches, though usually a little less. The important thing is that they be twice as long as they are wide to take full advantage of pattern possibilities. Regular bricks without holes aren't very suitable for mortarless paving. They are dimensioned to accommodate mortar joints and don't fit up well into many bonds without mortar.

Dig out an area about 4 inches deep as the first step in making a brick patio.

Block patio is laid in sand without mortar. Three sizes of units may be used.

Want something different? Cast your own units 1½-inch thick. Lay them in sand.

It should be smooth and as even as you can get it, yet should slope away from the house ⅛ to ¼ inch per foot. Fill any low spots with well tamped earth.

Frame two adjoining edges of the patio with bricks set on edge in rows. Place these in a trench dug 2 inches deeper than the rest of the excavation.

Spread 1 inch of sand and smooth it out. Roll out a layer of 15-pound asphalt roofing felt over the sand. This helps to keep the sand level and discourages plant growth through your patio. No guarantees.

Begin laying paver bricks at the corner where the edging has been placed. See some of the possibilities for patterns in the illustrations. When the whole patio has been paved, edge its other two sides with bricks set on edge.

A low-cost concrete curb can serve as an alternate edging for a brick patio. So can a well-staked wood form. Each is left in place. Even well established turf can serve as an edging.

CONCRETE MASONRY PATIO

Making a patio is easy and inexpenive with concrete masonry units. With a wide variety of pattern possibilities, you can make the design as simple or as wild as you wish. A few suggested patterns are shown in a drawing in this chapter. Some units come in colors.

The three most commonly used styles of concrete block for paving are the ordinary 4x8x16-inch hollow-core unit, the 2x8x16 inch solid unit and the 2½x8x16 solid unit. Sizes are nominal.

EXAMPLES OF PATIO BLOCK LAYOUT PATTERNS

Concrete masonry units for paving are laid in 2 inches of sand. When you excavate for the patio, dig deep enough to accommodate the sand fill. The blocks can be set in mortar on a concrete slab if you want a perfectly even surface that will stay that way.

Dig a trench along one edge of the patio sloping away from the house 1 inch in 15 feet. Set a row of blocks on edge in it to the grade of your lawn. You can also use 1x4 wood edge strips fastened to 2x2 stakes about 4 feet apart. Redwood is good because it resists rotting.

Notch a 1x6 board at its ends to the same thickness as a block (see drawing) and use this to level the sand. Rest one edge on a temporary screed placed board-length out from your new edging. You may have to level your sand subbase in several strips.

Start in one corner and lay blocks following a planned-on-paper pattern. Having the pattern to follow will save corrections while laying pavers. Shift blocks until they line up perfectly. Use a mason's trowel or sharp stick to jockey them around. If a block rests too high, lift it out and remove a little sand from beneath it. If it rests too low, remove it and add some sand under it. Then replace the block.

Lay the balance of the edging. Complete your job by sprinkling loose sand over the entire patio. Sweep it into cracks between the blocks, then hose off the excess sand with the spray from a garden hose.

Flagstone patios can be laid in sand just like those of concrete masonry. A better job, more in keeping with the high cost of flags is one built on a 4-inch concrete slab. Each flag is set in mortar placed on top of the concrete. The mortar bonds both flagstones and concrete together into a solid mass that will not heave with frost. Control joints in the base slab should be carried through into the flagstone pattern. If you don't do this, the flagstones will crack over every control joint. Some stones may even get torn loose by the action.

Further details on making flagstone paving are given in the chapter on beauty in stone.

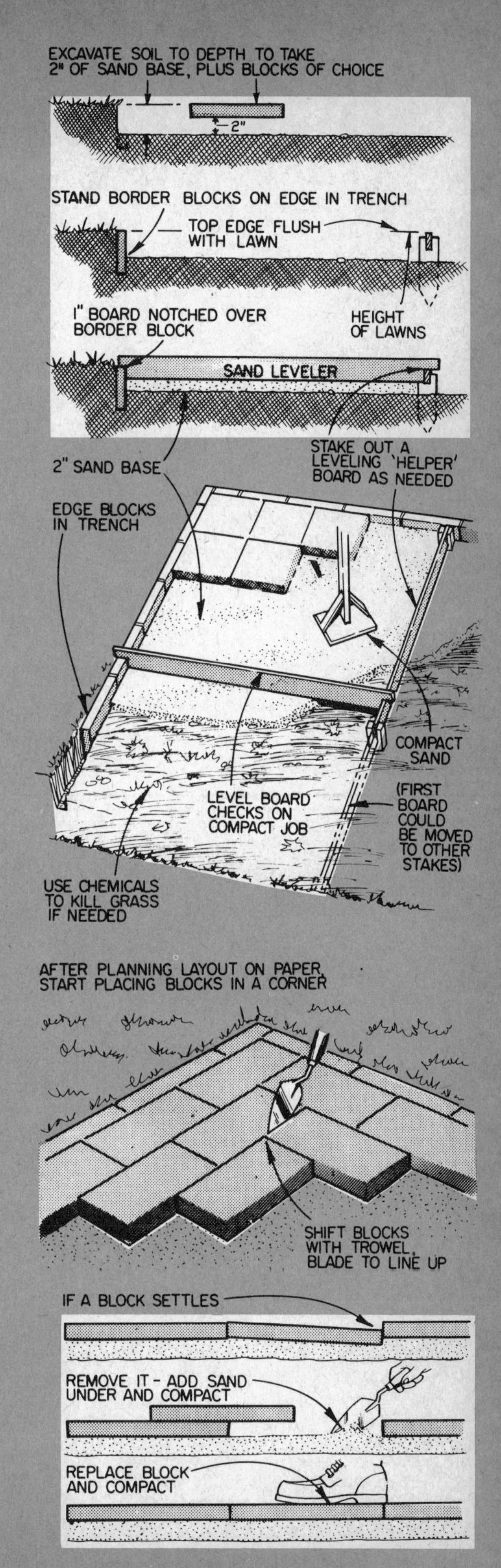

HOW TO MAKE SIDEWALKS

For safety, plan your driveway for a "turn-around" area to

Portland Cement Assn.

A boldly sweeping driveway of pebble concrete can be the focal point of your house.

A good looking sidewalk and driveway gives a finished, "dressed up" look to a house and grounds. Both are a welcome mat leading the eye from street to your front door. You can have whatever paving you want by building it yourself. Of the paving types, concrete is the lowest in cost, block next, then brick. Flagstone costs by far the most.

A driveway provides a hard-surfaced place for children to play off the street. The neat, crisp edges of a paved drive

AND DRIVEWAYS

make it possible to enter and leave your home with forward drive

add to the well groomed appearance of any yard. What's more, a driveway is your own private parking lot, even though it may only have room for one set of wheels. It's a place to clean your car, wash storm windows and do other tasks requiring a paved surface.

Besides its ordinary uses, a sidewalk serves children for recreation, things like rollerskating and hop-scotch.

The minimum driveway width for a one-car garage should be 8 feet. If the driveway curves, figure on at least 10 feet in width. The extra width should keep most drivers from straying onto your lawn. Double the width of the driveway for a two-car garage or carport. If the distance from street to house is great, a single drive may be built to within a couple car lengths of the garage then widened to some 18 to 20 feet. This permits a car to swing easily into either stall of the garage.

You can save a little money by using twin paved strips instead of a full-width driveway. If there are turns in the driveway, the strips should be even wider to keep cars from going off. The area between strips can be planted with grass, gravel-filled or covered with bricks or patio blocks. If you cover it with a hard surfacing, however, you're right back to the cost of a full-width paved driveway. Might as well go the whole route to begin with.

A driveway should not slope more than 1¾ inch vertically for every foot of horizontal distance. Changes in grade, either at the street or between, should be made as gradual as possible. Otherwise long-overhanging car tail ends may drag the drive as they pass.

A driveway that is dead level from street to house should slope from side to side at least ⅛ inch per foot of width. The cross-slope may be in the form of a

Concrete block pavers laid in a basket- weave pattern make steps and entry walk.

Robert Cleveland photo

Having more than a minimum driveway and parking area bids friends a warm welcome.

crown (higher in the center than at the edges) or it may slope all across (one edge lower than the other).

THICKNESS

Concrete is the most commonly used material for driveway-building. With all the color and texture possibilities it offers, there's no problem in getting the looks you want.

Knowing a few facts about concrete driveway thickness can save you money. If you build your driveway of quality concrete, you can safely shave its thickness to the bare minimum. Too many driveways are made 6 inches thick of sloppy concrete when 4 inches of quality concrete would do a better job. Make the decision for quality before you start. If only automobiles or light trucks will use your driveway, build it 4 inches thick. If delivery trucks will use it occasionally, build the drive 5 inches thick. If heavy fuel oil trucks or garbage trucks (they're heavy) use your driveway, you'd best go 6 inches thick. These are nominal thickness using 2x4 or 2x6 wood forms. To make a 5-inch driveway, use 2x4 forms and dig out the subgrade an inch below the form bottoms. You can get away with a 4-inch driveway that is really only 3⅝ inches thick. Don't go

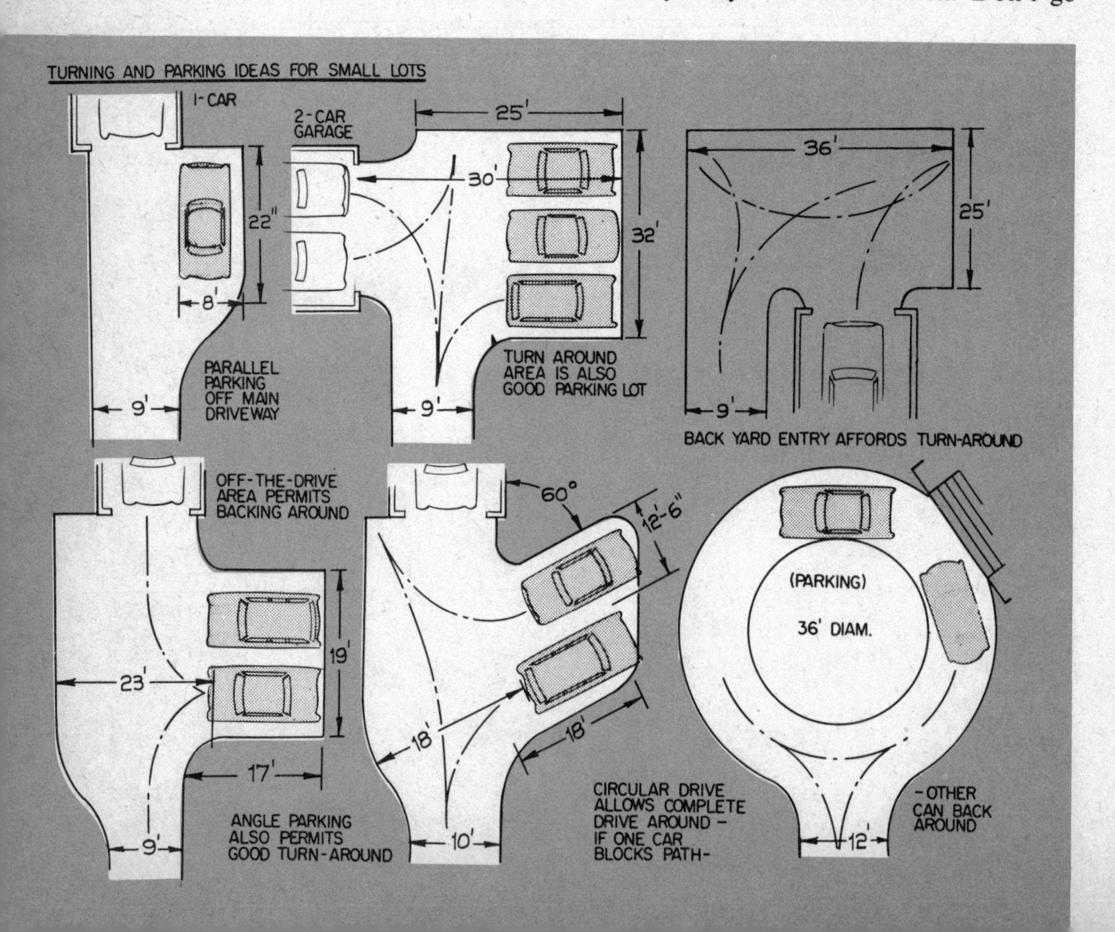

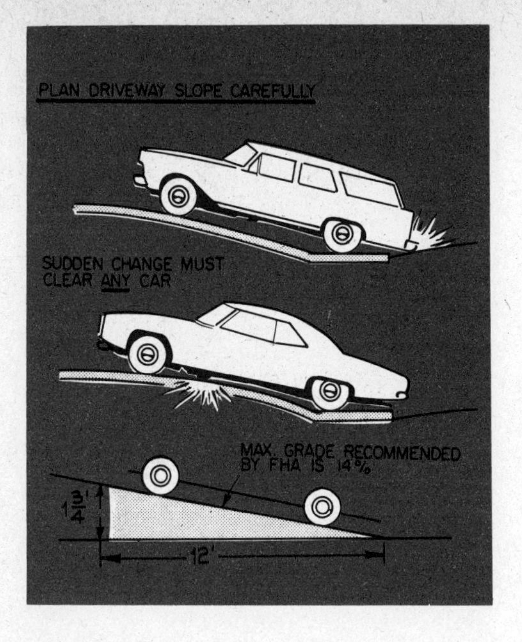

thinner in any spot or you're in trouble. If the driveway cracks under a load, all you can do is cry a lot.

After the forms have been set and staked, pull a template over them to check for subgrade depth. If you use a crushed stone subbase over a poor soil, the template will help you level your subbase grade.

To figure the amount of concrete you need for a driveway, see the accompanying table. Use only air-entrained concrete.

Don't let anyone talk you into putting "chicken wire" into your driveway slab. To be truly benefited you'd need much more steel than is in a little wire mesh. You're much better off to increase the thickness an inch and pay for the extra concrete with money saved on mesh.

Prepare the subgrade and form the driveway as described for concrete patio construction. A driveway is a kind of patio for your car. On vertical curves, where the driveway changes slope, use short lengths of lumber for forming. Tie a stringline to temporary stakes to determine the shape of the curve. Then lay your forms to this line. You'll get a smoother, more uniform grade change.

FORMING CURBS

Integral curbs can be made at the edge of a concrete driveway by using forms that are 4 inches higher than the slab thickness. Cut a ¾-inch plywood template to the shape you want for the driveway surface, including curbs and crown. Make it about 2 inches shorter than the distance between curbs so you can work it back and forth. Place a little extra concrete along the forms on both sides to provide material for the curbs. The strike-off template will help shape them. Additional curb shaping is done with a 3-foot length of 2×4. Draw it up the curb's curve holding it lengthwise of the driveway. Final shaping and smoothing of curbs is best done with float and trowel.

A curb should be about 4 inches high and 6 inches wide to the point where it flares into the driveway. Not only do curbs strengthen the driveway edge, they keep runoff water on the pavement. And they remind drivers to do the same.

Make control joints at 10-foot intervals along the driveway. A driveway wider than 13 feet needs an additional control joint dividing it. Often this joint is cut down the center of the drive, although it needn't be centered.

Provide full-depth isolation joints at sidewalks, street, garage and other slabs and walls.

Don't try to finish a driveway too smoothly. A wood or metal float finish is fine, so is a broomed finish. Cure for six days.

Concrete isn't the only material for building good driveways. You can use bricks, blocks or even flagstones if you like. The drawing shows how to make a heavy duty brick driveway using bricks placed on edge rather than flat. The different appearance will set your driveway apart.

DRIVEWAY ACCESSORIES

You can add such driveway accessories as spaces for backing, turning and parking. Don't make these smaller than recognized minimums for them or you'll

Sectioned forms are used to build a brick and concrete driveway shown on next page.

cuss every time you or someone else runs off the edge. A minimum turn-around pad is 20×30 feet. This is bare bones. It's better to go bigger. If there's room, it's best to provide for turning and parking with a projection off to one side of the driveway. Place it near the garage if you can with its sides flared into the driveway. Off-drive parking for two cars is created.

A similar turn-around pad coming off the driveway at right angles to the garage can be made large enough to park three cars. When empty, it makes turning around still easier.

If you have lots of room, a cul-de-sac driveway may be best. A 12-foot-wide paved circle lets you and guests wheel up to the door and drive out again without backing. A 60-foot circular area is required. The circle drive can be flared into a straight driveway. A drawback, one parked car blocks up the whole works. Shaving off and paving part of the center island can make room to stash that troublesome car out of the way.

Keep all plantings 2 feet back from the

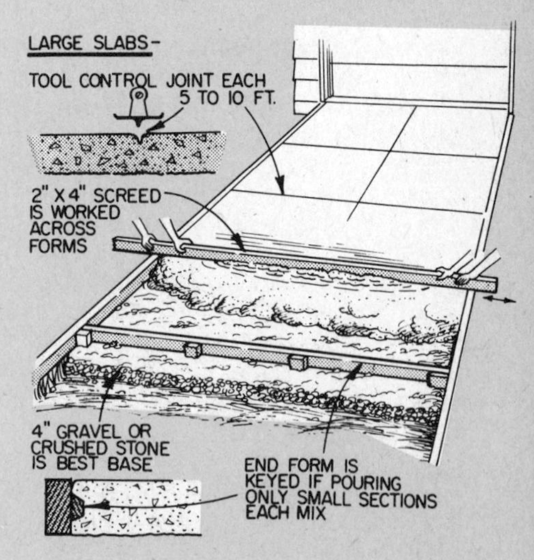

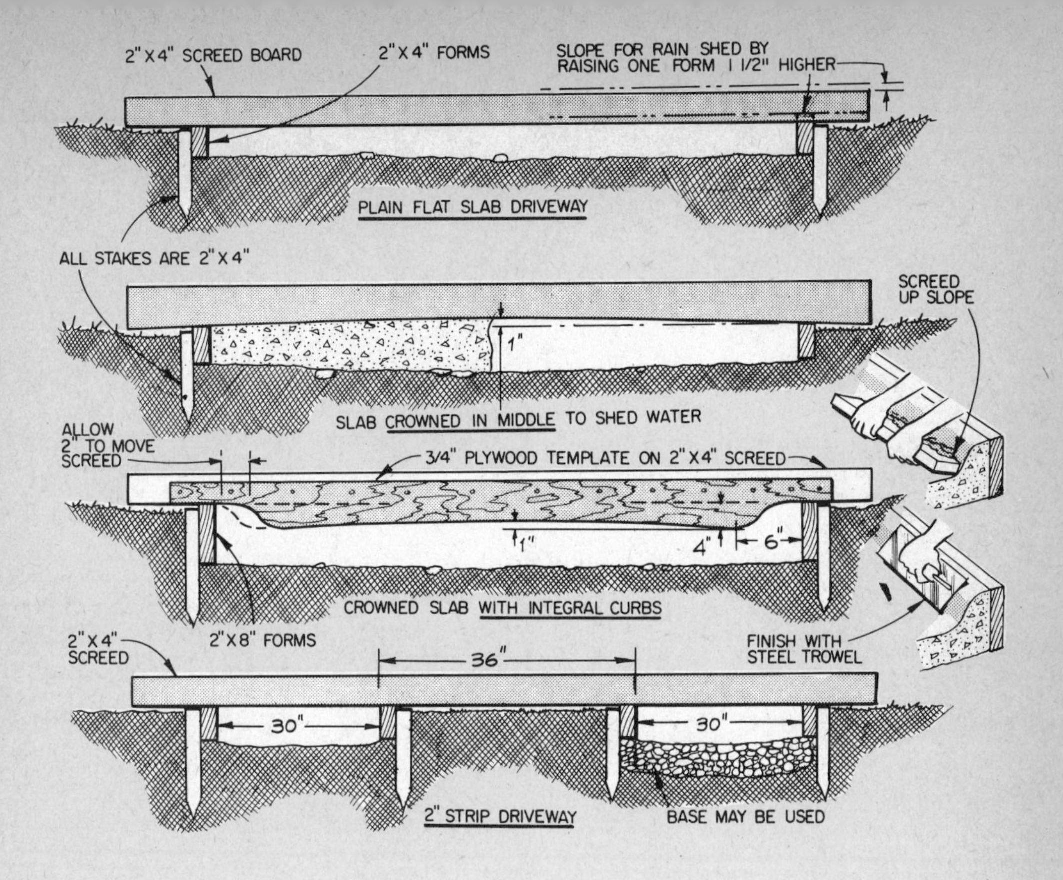

Finished driveway combines strips of brick pavers with texture of exposed-ag concrete.

Curving driveway avoids removal of tree and flares into the street for easy entry.

CONCRETE REQUIRED FOR SIDEWALK OR DRIVEWAY

	Area in Square Feet (width x length)					
Thickness	10	25	50	100	200	300
	cubic yards					
4 in.	0.12	0.31	0.62	1.20	2.47	3.70
5 in.	0.16	0.39	0.77	1.54	3.10	4.64
6 in.	0.18	0.45	0.90	1.80	3.60	5.40

Example: To find cubic yards of concrete needed for a 4-inch-thick, 9x45-foot driveway, first calculate the number of square feet. Multiplying 9 by 45 gives 405 square feet. This can be rounded off to 410 square feet.

From the table: 300 sq. ft. = 3.70 cu. yd.
100 sq. ft. = 1.20 vu. yd.
10 sq. ft. = .12 cu. yd.

Total: 410 sq. ft. = 5.02 cu. yd.

edges of curves so that cars can overhang the drive at these spots. Protect trees and shrubs with curbs or low walls.

PARKING

To get space just for parking, several tacks can be taken. One is to widen the driveway on one or both sides. For parallel parking the spaces should be 8 feet wide and 22 feet long to permit jockeying in and out. The driveway at that point should be 12 feet wide, making a total width of 20 feet.

Angle-parking uses less space. The ideal is a 60-degree angle. Parking stalls should be at least 12 feet 6 inches wide,

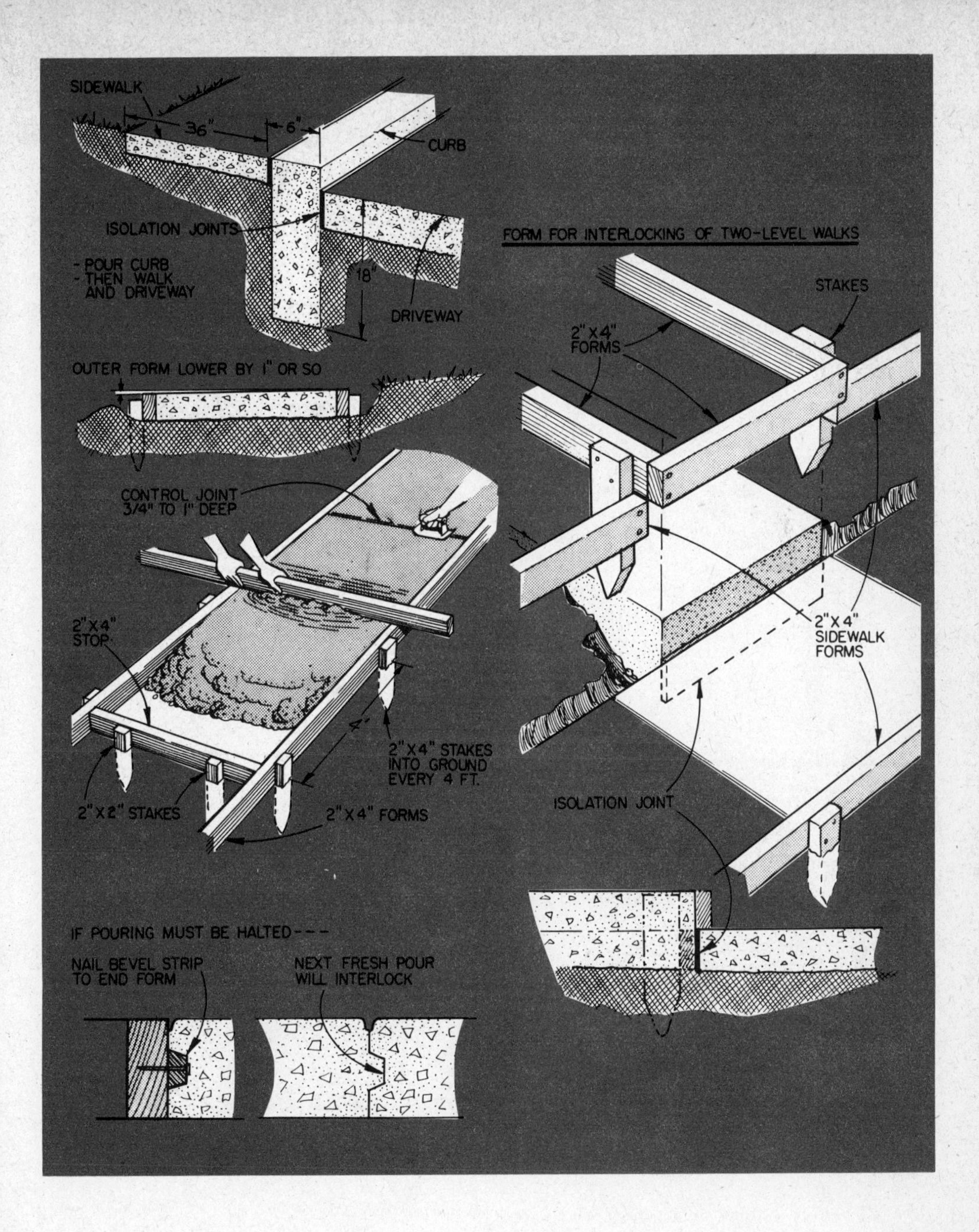

better yet 14 feet wide. This is measured at right angles to the car. You need an 18-foot length for the car plus another 18 feet behind it for maneuvering.

Straight-in parking stalls need a 9-foot minimum width, 17 feet of parking length and 23 feet more for backing out.

Provide curbs, cast-in-place or precast concrete tire bumpers or wood bumpers to keep parking cars from trespassing onto your lawn. Allow for 2 feet of overhang beyond the bumpers.

New sidewalk gets away from inconvenience of using driveway for access to side entrance.

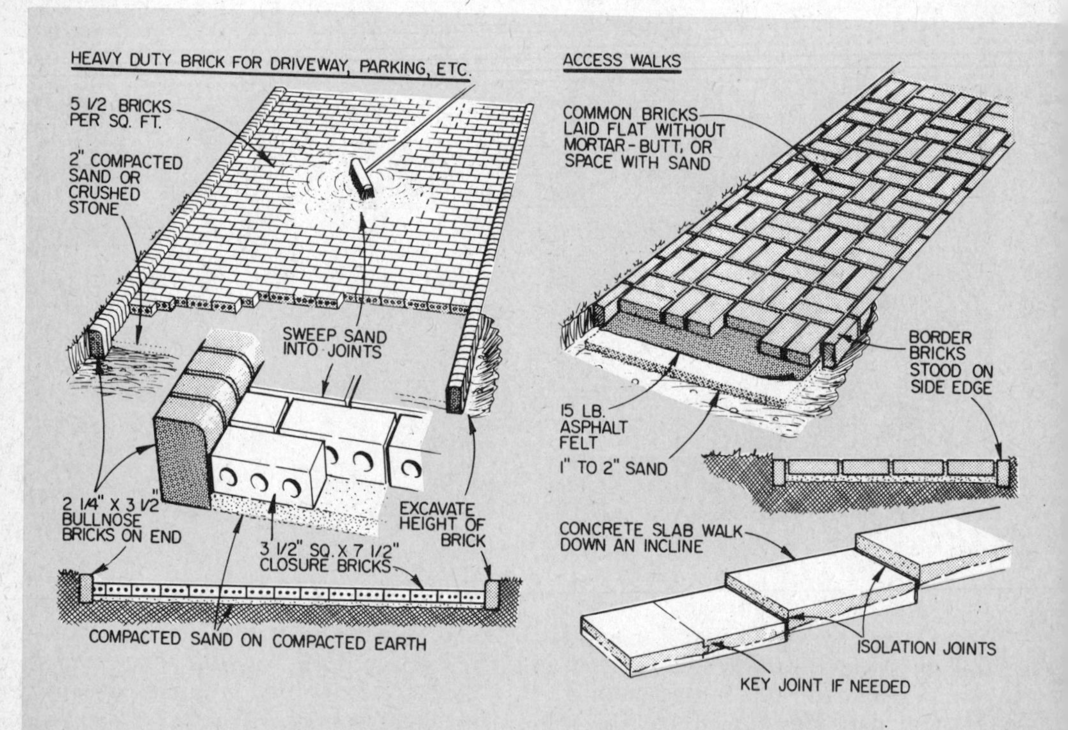

Broomed finish is best for driveways and walks because of nonslip gritty texture.

Even if you already have a driveway and sidewalk, you can add to them. The common driveway with 3-foot sidewalk running next to it can be improved by paving the area between walk and drive. Adding paved or graveled strips to one or both sides of a narrow access walk can effect an improvement. Paving a parking apron onto the driveway can, too. Double-strip driveways can be made better by paving them full width.

SIDEWALK TIPS

Walkways between your front sidewalk—the one provided by the city—and your house, house and garage, house and back yard, and other locations can be made of concrete, concrete blocks, bricks, adobe blocks or flagstones. Think of a sidewalk as a stretched-out patio. The methods already described for patio-building apply to sidewalks as well.

A sidewalk can be any shape. It can be finished in colors and textures. It can run continuously or be in sections, stepping-stone style. Main walks serving the front door should be 3 feet wide. Service walks need be only 2 feet wide. Sidewalk thickness should be a nominal 4 inches. Where trucks will drive across them, sidewalks must be 5 or 6 inches thick. Design them as driveways at those points. Control joints should be located every 4 to 5 feet along the sidewalk. There should be isolation joints between different thicknesses of sidewalk as well as at other slabs and walls. No reinforcement is needed. Make sidewalks of quality air-entrained concrete. Either finish the surface with a crown or build one form slightly higher to give slope for drainage.

Wherever you have to stop work on the sidewalk, put a wood bulkhead in the form. Make it from a length of 2×4 cut to fit between the forms. Nail a beveled 1×2 along the bulkhead to form a keyed construction joint.

Sidewalks are often finished with a wood or lightweight metal float, then allowed to set up. This imparts a gritty nonslip surface texture that helps to prevent falls.

RESURFACE AND PATCH

Make repairs early or you will have a major job to do

Got a driveway that's too low to meet the garage floor? Or a patio full of "bird-baths?" How about a cracked step or a scaled sidewalk? There's no need to live with these problems. You can either hide them under a new layer of concrete or patch them. Which you do depends on how extensive the damage is and whether they're slabs or walls. You can resurface a slab but not a wall. A cracked wall can only be patched. Resurfacing is used to renew, cover up or raise an existing concrete slab. It also can beautify, as in resurfacing with a colored concrete or rustic terrazzo topping.

Concrete is resurfaced in one of two ways: bonded or unbonded. Bonded resurfacing is stuck strongly to the top of the old slab and can be as thin as ⅝ inch, even thinner in some cases.

Building an unbonded resurfacing is much like casting a whole new slab on top of the old one, using the old slab merely as a subbase. The resurfacing must be at least 2 inches thick and should be reinforced with steel mesh weighing at least 30 pounds per hundred square feet. The mesh should be placed in the middle of the resurfacing, a good trick if you can do it. A layer of sand or a polyethylene bondbreaker is spread over the base slab before placing the resurfacing concrete. It keeps the two from bonding into one structural unit. Control joints and isolation joints in the new slab are made according to the rules given in another section. An unbonded resurfacing should be used where a raise in the elevation is desirable, or where the old slab is badly cracked.

BONDED RESURFACING

To take a bonded resurfacing, the existing slab must be structurally sound. If it's cracked, the resurfacing will crack, too, within a short time.

To get a good bond for resurfacing, first chip the old slab's surface with a mason's hammer. Brush off all the loose particles. Thoroughly clean the surface of grease, oil, paint, wax and other materials. Grease and oil can be removed by scrubbing with trisodium phosphate (TSP) sprinkled onto hot water poured over the surface. Paint must be chipped or ground off.

Provide "tooth" to the old surface by etching with muriatic acid, one part acid in five parts water. Flush the surface to remove all traces of acid. Keep the slab wet overnight. Before placing your resurfacing remove all free water on the surface.

BONDING

You have your choice of two methods of bonding a resurfacing: portland cement slurry bond, and bonding agent. The cement slurry bond is cheaper but requires the chipping and acid-etch previously described. To use it, sprinkle portland cement over the wetted slab and scrub it into a thick slush coat. Brush this out well to avoid a too-heavy coating. Resurfacing concrete is placed right over the slush coat before it dries and whitens. If it whitens, its bonding properties are lost.

When using a bonding agent, such as *Weld-Crete*, you can omit the chipping and acid-etch. Tightly clinging paint need not be removed. Loose paint should be taken off. Bonding agent comes in ½ pints, pints, quarts and gallons and should be available through your building materials dealer. Material cost is about 3 cents a square foot or more, depending on the size of the job. Big jobs lower the per-square-foot cost. The agent may be applied by brush, roller or spray. Resurfacing concrete is placed on top of the bonding agent after an hour of drying.

Bonding agent makes the job of patching and resurfacing much easier. Just brush paste on and patch.

Larsen Products

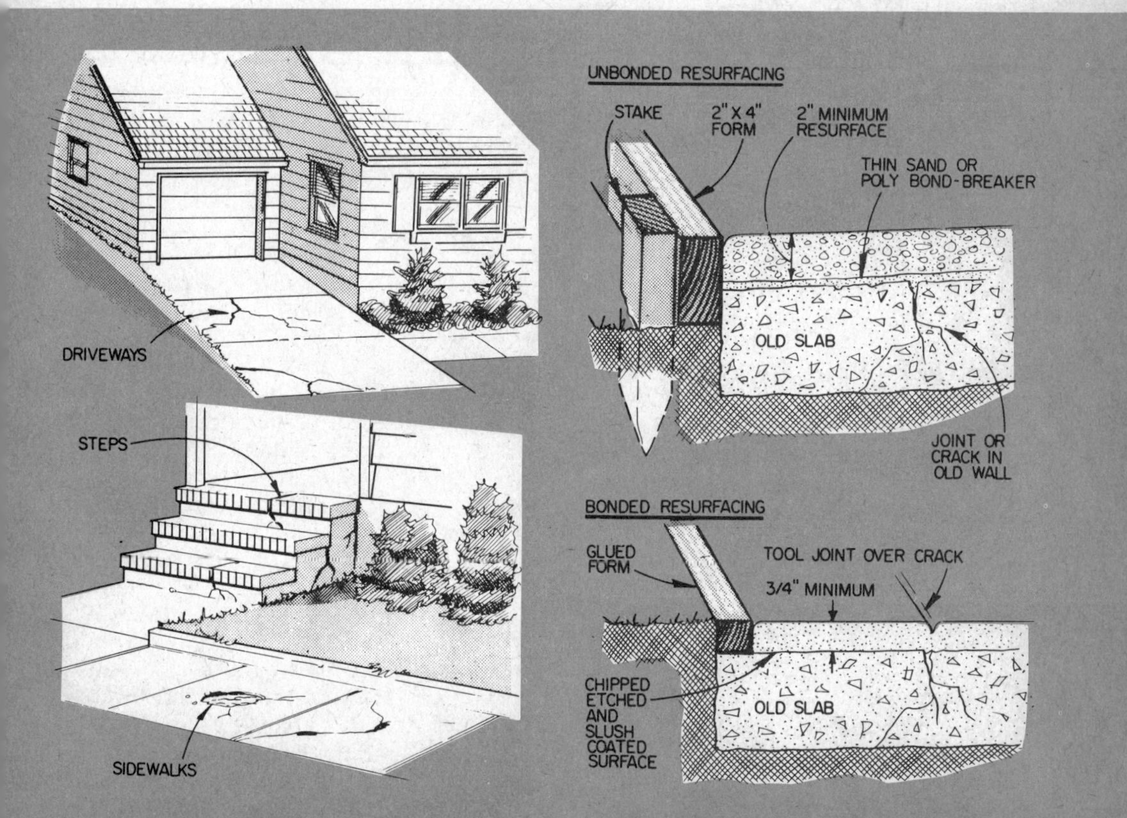

For a bonded resurfacing adhere 1″ x 1″ strips to old slab with paneling adhesive.

Sprinkle cement on wet slab and brush it into a thick slush coat to create a bond.

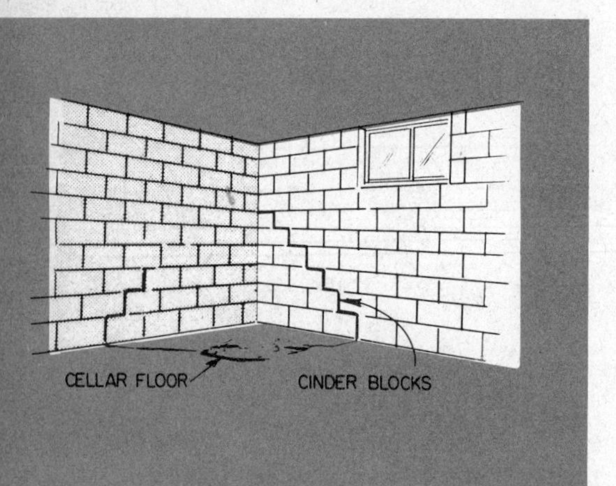

To use special concrete patch, add water and mix. Let stand 10 minutes and mix again. Apply the patching compound to the cleaned patch, level it and finish smooth.

If a thin resurfacing is wanted, such as one that will trowel out to a feather-edge, bonding agent can be applied both to the surface and as an additive in the resurfacing mix water. Make the mix water of one part *Weld-Crete* in three parts plain water. Used this way, bonding agent permits feather-edging, according to the manufacturer.

RESURFACING MIX

For resurfacing, use a concrete mix of 1 part portland cement, 1 part sand and from 1½ to 2 parts coarse material, such as pea gravel. Parts are by volume. Use as little water as possible – not more than the equivalent of 5 gallons per bag of cement. Adjust the amount of sand to get the desired workability. The mix should be air-entrained, both for easy working and for frost-resistance. If the resurfacing is to be feather-edged, better use fine sand to make the mix.

Place the concrete, finish it with a wood float, then steel-trowel if a slick finish is wanted. Start the cure as soon as you can without marring the surface and keep it up for at least 6 days.

Forms should give your resurfacing the exact grade wanted. If the slope of

Before slush coat dries and turns white place your own mix, or ready-mix topping.

Strike off resurfacing across tops of the forms. Edge, joint and finish as usual.

the old slab isn't what you want, you'll have to make the forms higher on one side than the other. If the old slab is properly sloped, you can simply fasten 1×1's or 1×2's laid flat directly to it. A good way to secure these is with panel adhesive. It comes in drop-in cartridges for your calking gun. Apply a continuous ⅜-inch bead of adhesive down one side of the strip. Contact the strip where you want it, then pull it away. Allow from 1 to 10 minutes of air-drying, then recontact the strip. Walk on it to adhere it firmly. The best working adhesive, I found, were U.S. Plywood's *Panel Adhesive*, flammable type, and Masonite's *Royalcote* adhesive.

If you use redwood forms, they can be left in place. Stay-in forms may be run anywhere they're needed, even along cracks. Take-off forms should be run around the outside edges only.

Make marks on the forms indicating where all joints and cracks in the old slab fall. These will have to be cut into the resurfacing with a grooving tool. Neglect them and you'll get ugly cracks. No matter how it looks, a joint must follow exactly every crack in the old slab, hairline cracks excepted.

If you do the resurfacing properly, there should be no unbonded spots. Faulty bond shows up when the hardened resurfacing is tapped lightly with a short length of metal rod. Unbonded spots will sound hollow.

PATCHING

A patch is simply a small, thin featheredged resurfacing. You may use one of the proprietary patching compounds such as *Coughlan Concrete Patch*, *Aqua-Dri Plus* or the *Weld-Crete* system just described for resurfacing. Or you can make your own patching mix using the following formula: Add 2 pounds of plaster of Paris to 45 pounds of ready packaged sand mix concrete. Mix dry. To use, add ice water and patch quickly on top of a cement slurry or bonding agent treatment. You needn't wait long for the patch to set up. The plaster of Paris kicks off the reaction quickly.

Finish your patches to look as much like the surrounding concrete as possible. Don't expect a perfect color match. Close is all you can expect.

Patching and resurfacing sure beat breaking up an old slab and building a whole new one.

PRECASTING BEAUTY IN CONCRETE

Functional and decorative objects can be made indoors

When the weather isn't suitable for concrete or masonry work, there's no need to hang up your tools. Put them to work on fun-to-make precast projects. Working in your garage or basement, you can build things like birdbaths, planters, wall plaques and patio tiles that will make you feel like a prolific sculptor. In precasting, the object is made at your convenience, then moved where it's wanted.

Basic to all precasting is the form or mold. The most common form material is wood, though you're by no means limited to it. Patio tiles can be cast in a 1×2 wood form on a plywood or hardboard base. Larger projects, such as fence parts or a picnic table, call for 2×4 forms. Still others need ½- or ¾-inch plywood box-shaped forms that are reinforced around the edges with 1×2 or 2×2 wood members.

Before building a wood form, plan ahead for form removal after the concrete has set. If you're casting a one-of-its-kind project, the form can be decimated during removal. For multiple-use you'll have to build it with screws, hinged corners or protruding nails to permit nondestruct take-apart.

Another method of getting the form off is to design it with *draft*, that is, sloped sides so the precast item can be slipped out easily. In any case the form should be well oiled with old crankcase oil or cheap nondetergent motor oil to let it come away from the concrete with ease. Oil after each use.

Wood forms can impart surface textures other than that of smooth-sanded wood. Some of the hardboards and exterior plywood siding give interesting textures. Casting concrete in a pegboard form gives it a pimply-dimply effect. Take such a form off before the concrete gets too hard.

Other materials can be fastened inside a wood form to give texture. Rubber matting and formed plastic are made for this purpose, but not generally available to us handymen. Rubber stair tread materials will work nicely, however. So will carpet-like rubber mats.

Boston Woven Hose Div. of the American Biltrite Rubber Co., Inc., Box 1071, Boston, Mass. 02103, produces rubber matting for texturing concrete. Monsanto Chemical Co., Plastics Division, Springfield, Mass., is a supplier of plastic form liners.

METAL MOLDS

You can buy prefabricated metal molds—usually aluminum—for making a wide variety of lawn and garden items. With them you can cast such things as lawn deer, benches, small figures, garden fountains, birdbaths and such. In order to recoup what you put out for molds, you have to sell some of the products or else find a buyer for your used mold. Metal molds can be used over and over again. The pieces unbolt to lift off, even the most involved designs. One source for aluminum molds is Concrete Machinery Co., Inc., Hickory, N. C. 28601. This firm also sells a book with detailed instructions on all phases of precasting in metal molds.

An attractive driveway can be made with cast concrete pavers, grass between them.

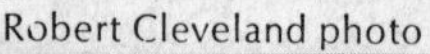
Robert Cleveland photo

Robert Cleveland photo

Precast concrete fence parts were built in winter, set up in spring. Slots in the sides of the posts take angling rails. The posts are set in concrete.

Wood forms saber-sawed and joined to make a curving design are filled with concrete.

Using a dry concrete mix, the forms can be lifted away as soon as the tile is finished.

Plywood form for a concrete planter takes apart and lifts away from hardened planter.

Besides the conventional forms, molds can be made from whatever is at hand. Gallon milk cartons or plastic gallon bottles are just right for flowerpots. Make depressions for the flowers by shoving a half-gallon carton or a can down into the fresh mix. Old 10-quart oil cans make larger rectangular pots. And the mold is already oiled.

Corrugated cardboard shipping cartons (see your grocer) make good forms for larger planters. Use a large carton for forming the outside of the mold and a smaller one weighted for the inside.

You can dig a hole in the ground, line it with polyethylene sheeting or paper and cast concrete in it. Huge tubs and planters can be made this way. Just be sure you can lift them out. The inside of such a mold can be formed with plywood pushed down into the concrete mix and weighted to stay.

Almost anything you can dump concrete into can be used as a mold. An old tire, a hose laid out on the plastic-covered driveway, a water pail, all can serve as molds.

MIXES

No matter what form you cast in, your selection of mixes is wide open. The table shows how to proportion them.

General-purpose gravel-mix – For items with large cross sections, like fence posts, this is the best mix. It gives a plain, smooth concrete finish with the same texture as the form. The largest aggregate in the mix should be less than one-fifth the thinnest cross section of the item being cast. No different than any con-

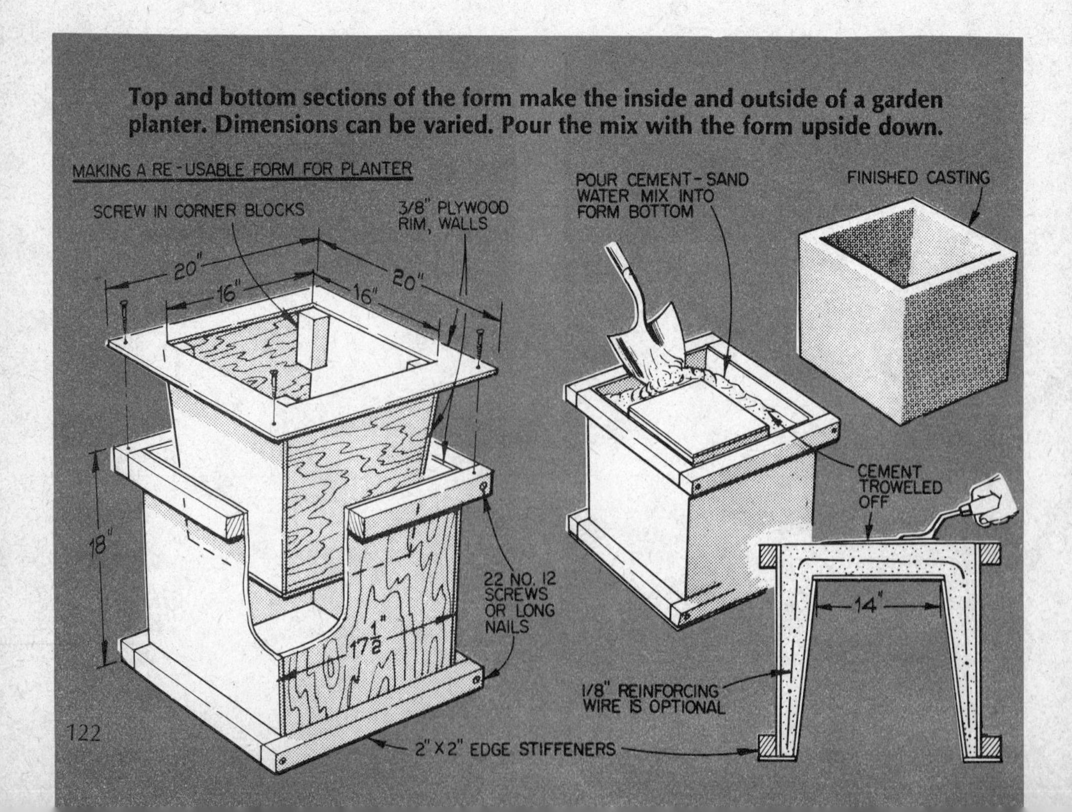

Top and bottom sections of the form make the inside and outside of a garden planter. Dimensions can be varied. Pour the mix with the form upside down.

crete. Ready-packaged gravel-mixes can be used if they meet this requirement.

Sand-mix—Precast items with thin sections should be made of sand-mix concrete. It, too, comes ready-packaged. Sand-mix finishes to a normal gray color, like gravel-mix. Sand-mix can be made into a "pourable" mix doubling the normal water or into a "super-pourable" mix by using twice as much water as cement. A "pourable" mix is best for picking up fine detail in a mold where maximum strength is not required. A "super-pourable" mix is used in a metal mold to eliminate air bubbles in the casting and get a very tight, closed finish. Strength is thrown to the wind. Excess water that works to the top of the mold should be poured off and replaced with mix.

Hand-packed mix—This mix takes on a uniform sandy finish that looks great. With the tamping, it's more work than pouring a mix, but strength isn't sacrificed. The finer the sand you use, the smoother the finish. For hand-packing, the mold must be designed so that all portions of it can be reached for tamping through the bottom opening. In most cases the items are so strong you can strip the form right after casting. There's no need to oil the mold. It won't stick.

A good hand-packed mix starts with a general purpose sand-mix using fine mortar sand. You may add some small-size gravel to the mix if you like. You may make it from ready-packaged sand-mix using the amount of water shown in the table. If you're proportioning your own, and using wet sand, don't add *any* water to the mix. The sand already contains enough. With drier sand that's only damp, you may need to add about one-eighth part water. 'Taint much. Add water only after the sand and cement have been blended thoroughly. The mix should squeeze like brown sugar in your hand without any water squeezing out of it. Very wet sand contains too much water for use in a hand-packed mix. See the definitions of "damp," "wet" and "very wet" in the first chapter.

Vermiculite mix—Vermiculite is expanded mica, normally used for loose-fill house insulation and as a garden

Pouring method with a commercially made metal form ensures bubble-free castings. Pour full, tap sides and turn if needed.

Insert a rod for bottom drain hole and strike off the mix. Pour off any water that floats up and replace with concrete.

When precast object has hardened, mold can be taken apart and stripped gently. Core lifts out, sides unbolt and come apart.

Concrete Machinery Co.

The hand-packing method uses almost-dry sand mix and makes strong, sand-textured objects. Tamp in one- to three-inch lifts.

Concrete Machinery Co.

Mold may be removed immediately, leaving the object able to stand on its own. You can get real production in this method.

The exposed-aggregate bicycle caddy was cast in retarder-coated wood form using white stones and white concrete. The aggregate was exposed by wire-brushing after two days.

Robert Cleveland photo

Precast concrete urns can do a lot for your house when filled and placed prominently.

mulch. It makes good lightweight concrete for nonstructural uses such as wall plaques. Vermiculite concrete castings can be carved, sawed and drilled. Just like wood, but no grain. Use them indoors only. Sand in the vermiculite mix makes it heavier, stronger and able to reproduce finer detail. White cement used with vermiculite makes a sandy-colored surface that is somewhat lumpy.

Lightweight aggregates—Materials such as Haydite, Perlite, pumice, scoria and crushed lava make a lightweight concrete with some structural properties. This material can be used outdoors. If weight is a problem and you can get a good lightweight ag, don't hestitate.

No-fines mix—No-fines concrete is made with cement and coarse aggregate but no sand. It is lighter and not as strong as a mix with sand, but is terrific for non-load-bearing precase items. The surface is open.

No-fines concrete requires just the right amount of water. Too much makes a watery cement paste that won't stick to the aggregate. Too little produces a paste that won't coat every stone. Add water in the mixer a little at a time. Stop adding when all the stones have become coated with cement paste. The amount of aggre-

HANDY MIXES FOR PRECAST CONCRETE (parts by volume)

TYPE	Cement	Sand	Coarse Aggregate	Water (Average wet sand)	Properties
General purpose gravel mix	1	2½	3 (small stones)	½	Hard, structural concrete. Tap sides of mold to remove bubbles.
General purpose sand mix	1	3	4	½*	Hard structural concrete, but less yield than above.
Trial mix for hand-packed concrete	1	3	0	⅛ to ¼ (see text)	Hard structural concrete for hand-packing into mold.
Vermiculite aggregate	1	0	4 to 6	1½**	Lightest lightweight concrete. Low strength, poor weathering. Can be carved.
Vermiculite aggregate	1	2	3	1½**	Lightweight concrete, poor weathering. Use more cement for added detail.
Haydite, Perlite, pumice, crushed lava aggregates	1	2	3	½*	Lightweight concrete, good weathering, fair strength. Heavier than with vermiculite aggregates.
No-fines trial mix	1	0	9 (⅜" dia.)	½ or just enough to cover aggregate	Honeycombed with air spaces. Weighs about three-fourths as much as ordinary concrete.
Terrazzo Topping	1 (white)	0	1 (marble chips)	½	Looks like terrazzo when surface is brushed to expose marble chips.

* Use 1 part water to make a "pourable" mix
** Use 2½ parts water to make a "pourable" mix

Cast pea-gravel-mix in a gallon milk carton. In ten hours brush to expose texture. Can was used to make planter pot.

gate indicated in the table may have to be adjusted, too, to suit your materials. There should be just enough cement paste to completely coat and glue together all the stones. Not enough to fill between them.

Terrazzo topping mix—You can make precast items with a terrazzo-like surface by casting them of this mix. Brush the surface mortar away as soon as the form can be removed (about 8 to 10 hours). Color is usually added to the mix.

The terrazzo portion of a precast object need be only skin deep. Place it in the mold first, then mix up a general-purpose sand-mix and put that in the mold on top of the terrazzo. The two materials will bond perfectly while they're both fresh.

One way around the terrazzo exposing problem is to paint the form with a *re-*

Terrazzo tiles were cast face down, using two-course method, turned over after 8 hours and brushed to expose marble chips.

Sand-casting is fun. Tool impressions in sand remain after the tools are lifted. The form is made of 1x2's and hardboard.

To produce pebble concrete by sand-embedment sprinkle stones into a thin layer of damp sand and pour the concrete over them.

At left, completed sand casting is great for office or den wall. Screw-eyes embedded in back hold a wire for hanging.

Decorative patio tiles can be made by casting upside down in a form on wrinkled polyethylene sheeting to give an elephant-skin texture.

tarder before casting in it. The retarder holds back the set of surface concrete that touches it. Two days later when the object has set up, the surface layer will still be soft. You can strip the form and brush away surface mortar to expose the colorful marble chips beneath.

Retarder can be used for any concrete casting where surface exposure is wanted. You need the water-insoluble type. The heavier you apply it, the deeper its effect. If you can't get retarder locally through a concrete products supplier, write to Burke Concrete Accessories, Inc., 3870 Houston Avenue, San Diego, Calif. 92110. Ask about *Agreveal-F*. A gallon should last you a lifetime.

If you do much precasting, it will pay you to buy a small mixer. The handiest of these uses a 5-gallon pail as a mixing drum. It sells for about $40. Get it through the Montgomery Ward mail order catalog. Though small, the mixer will handle just about any precast project you could want. If one batch won't fill the form, fill it in a number of batches.

UNUSUAL TECHNIQUES

Plastic form liners mold precast concrete into all manner of interesting designs. A glossy surface quite unlike the usual concept of concrete, and more like china, results because the object takes on the plastic liner's smooth surface.

Another way to get a glossy surface is to precast concrete on glass. The object will be smooth and flat just like the glass. Be sure to vibrate it well to remove all air bubbles against the glass. One way to do this is to place an orbital sander next to the form for a few seconds. It's tough on sanders but it produces smooth-textured concrete.

For a purposely pock-marked texture, spread rock salt into the form and cast concrete over it. The rock salt will later wash out leaving holes. This treatment isn't recommended for outdoor precast objects that must stand up to wetting and freezing.

Polyethylene sheeting installed in a form gives a glossy texture to concrete cast against it. A variation is to wrinkle up the poly and cast concrete against the wrinkles letting them show on the finished surface. Another variation is to place poly over a rough surface, such as one strewn with pebbles. Cast concrete on top of that. When the polyethylene is peeled off, the rough cast surface is most striking. Tiles made this way are fine as stepping stones. Besides the different faces poly sheeting can give concrete, it makes a good parting material for castings. Concrete lifts away nicely. Interesting flat precastings can be make by sticking broken or whole ceramic tiles and other flat items face down in a form. Pour concrete around them. The finished casting strips from the form with the tiles exposed at its surface. Double-faced tape like that sold for laying carpet-tiles does the stick-down job.

INDEX

Index Key:
Chapter heads are in capital letters
Bold face numerals indicate a photo
Italicized numerals indicate diagram